D1154855

PRAISE FOR *ESCAPING GRAVITY*

"In *Escaping Gravity*, Lori Garver, career champion of everything space, offers a front-row seat to the decades-long struggles within and among space bureaucrats and space billionaires. From presidents to heads of agencies to astronauts to key members of Congress, she knew them all and they knew her—and they did not always see eye to eye. Bring popcorn as you bear witness to an untold slice of space history."

—Neil deGrasse Tyson, astrophysicist, American Museum of Natural History, and author of *Space Chronicles: Facing the Ultimate Frontier*

"Lori Garver is a self-proclaimed 'space pirate,' who took on NASA, the aerospace industry, and even members of Congress as part of a tumultuous groundswell that helped launch the modern commercial space industry. Told without fear or favor, her compelling tale transcends the space industry and shows us, from the inside, how Washington works—who wins, who loses, and who bears the scars."

—Christian Davenport, author of *The Space Barons: Elon Musk, Jeff Bezos, and the Quest to Colonize the Cosmos*

"As much as anyone not named Elon Musk, Lori Garver has helped prod, push, and pull NASA to embrace commercial space and all of its promise. Now, as this industry has become the envy of the world, Garver tells how it was done over the last three decades. She names names, yanks skeletons from closets, and writes illuminatingly about the brilliant future ahead of us."

—Eric Berger, Ars Technica reporter and author of *Liftoff: Elon Musk and the Desperate Early Days that Launched SpaceX*

"Garver was the clear administration leader for creating and sustaining commercial green shoots in space enterprise.... She [was] a loyal and effective advocate for Obama administration space policies and clearly had significant influence in administration space policy development."

—Mark Albrecht, former executive director of the National Space Council, George H. W. Bush administration

"Lori Garver was the catalyst that helped bring NASA into the new Space Age. No one has done more than Lori to stand up to parochial (and patriarchal) interests in this industry and usher in a more meaningful and sustainable space program. *Escaping Gravity* is an enduring story of how a woman with a different background and perspective can make a difference—and become a force to be reckoned with."

—Emily Calandrelli, award-winning science communicator and host of FOX's *Xploration Outer Space* and Netflix's *Emily's Lab*

"*Escaping Gravity* delivers the inside story of today's commercial spaceflight revolution. Before SpaceX, Blue Origin, Virgin, and XPRIZE, the space industry was an old boys club, with a handful of military defense companies. It was ultraexpensive and risk averse. Lori Garver led the charge to topple the old school and engage the entrepreneurial engine we now see today. If it weren't for Lori Garver's passion and vision, we would not be on our way to the Moon and Mars today."

—Peter H. Diamandis, MD, founder/chairman of XPRIZE and Singularity University and author of *Abundance*, *BOLD* and *The Future is Faster Than You Think*

"If you're interested in the story of a pioneering woman revolutionizing the male-dominated space industry, read this book. If you're interested in the key early turning points of Elon Musk's and Richard Branson's space efforts, read this book. If you're interested in how to affect substantive change within government, industry, and the nonprofit sector, read this book! Lori's life story is filled with incredible adventures, and she has had a profound impact on humanity's future in space and on Earth."

—George Whitesides, former CEO of Virgin Galactic

"Lori Garver's *Escaping Gravity* is a must-read to understand the twenty-first century remaking of the US space industry. Her clear-eyed view from inside NASA and the halls of American power reveals the why and how behind the new Space Age that's now underway. Lori artfully gives personal introductions to dozens of influential people—from the billionaires who are now household names to the politicians, officials, celebrities, entrepreneurs, executives, and more involved. Her story is gripping and personal, interwoven with never-before-told conversations, historical facts, and battles of will."

—Michael Sheetz, CNBC space reporter

ESCAPING GRAVITY

My Quest to Transform NASA and Launch a New Space Age

LORI GARVER

DIVERSION BOOKS

For David, Wesley, and Mitchell—the brightest stars in my universe.
And for all who light the path to a better future

Diversion Books
A division of Diversion Publishing Corp.
www.diversionbooks.com

First Diversion Books edition, June 2022
Hardcover ISBN: 9781635767704
eBook ISBN: 9781635767735

Printed in The United States of America
1 3 5 7 9 10 8 6 4 2

Library of Congress cataloging-in-publication data is available on file

CONTENTS

Look again at that dot. That's here. That's home. That's us. On it everyone you love, everyone you know, everyone you ever heard of, every human being who ever was, lived out their lives. The aggregate of our joy and suffering, thousands of confident religions, ideologies, and economic doctrines, every hunter and forager, every hero and coward, every creator and destroyer of civilization, every king and peasant, every young couple in love, every mother and father, hopeful child, inventor and explorer, every teacher of morals, every corrupt politician, every 'superstar,' every 'supreme leader,' every saint and sinner in the history of our species lived there—on a mote of dust suspended in a sunbeam . . .

Our posturings, our imagined self-importance, the delusion that we have some privileged position in the Universe, are challenged by this point of pale light. Our planet is a lonely speck in the great enveloping cosmic dark. In our obscurity, in all this vastness, there is no hint that help will come from elsewhere to save us from ourselves.

—CARL SAGAN, *Pale Blue Dot*

FOREWORD

THE SPACE AGE THAT BEGAN IN THE LATE 1950S HELPED TO DEFINE THE twentieth century. The realms above our Earth's atmosphere began to be populated with satellites and spacecraft that connected the world and revealed new knowledge of our planet and the universe. Early US exploits in human spaceflight, which sent the first humans to walk on the Moon, showed the world that we could meet bold objectives and accomplish feats previously considered unachievable.

In the class that I teach at Tulane University on the history of technology, the students discuss the drivers of innovation. Are big government projects more effective? Or are nimble entrepreneurs? The answer, of course, is that stunning breakthroughs usually involve a symbiotic mix. After being appointed by FDR in 1940 as head of the National Defense Research Committee, Vannevar Bush—dean of engineering at the Massachusetts Institute of Technology and cofounder of Raytheon—oversaw the government science programs that resulted in the atom bomb and electronic computer. In his seminal 1945 paper, "Science—the Endless Frontier," he described how a collaboration of academia, business, and government would drive innovation.

The Electronic Numerical Integrator and Computer (ENIAC) was funded by the United States Army at the University of Pennsylvania. It's an origin point for the Universal Automatic Computer (UNIVAC) and most other electronic computers. Today's internet can trace its lineage to the network funded by the Defense Advanced Research Projects Agency (DARPA). For the biotech industry, the dawn was the sequencing of the human genome, largely funded by the National Institutes of Health. Innovation after innovation have followed this collaborative path.

Lori Garver has been a brave and effective leader in making sure that America's space program followed this "innovation progression" that involves collaboration between government agencies, most notably NASA, where she served, and private companies, such as those led by Elon Musk, Jeff Bezos, and Richard Branson.

A year after launching SpaceX, Musk said, "Just as DARPA served as the initial impetus for the internet and underwrote a lot of the costs of developing the internet in the beginning, it may be the case that NASA has essentially done the same thing by spending the money to build...fundamental technologies. Once we can bring the sort of commercial, free enterprise sector into it, then we can see the dramatic acceleration that we saw in the internet."

Garver's drive to get NASA to cooperate with private companies was critical. After NASA succeeded in the race to land on the Moon, subsequent presidents made similar proclamations about returning humans to the Moon to establish bases and as waypoints to Mars. But NASA's proposed implementation of these programs came with Apollo-scale price tags, without Apollo-like justification.

The exorbitant institutional costs required to implement its large, centralized government-led missions stifled NASA's ability to keep innovating and lowering costs. Traditional programs established to replace the Space Shuttle all succumbed to the reverse incentives inherent in government contracting.

By the summer of 2008, when Garver was asked to review NASA for presidential candidate Barack Obama, NASA's Space Shuttle was scheduled to be retired in two years and its replacement program had veered so far off course that its only recourse was to pay the Russians, our former Cold War adversaries, to ferry its astronauts to and from the International Space Station. Even so, when the Obama administration recommended NASA turn to the US private sector to accomplish what it had not been able to do on its own, the idea was scorned by the NASA standard bearers, aerospace industry, and Congress.

Most of us are now well aware of the impact recent private sector investments have had on reducing the cost and increasing the capability of space activities. Entrepreneurial companies are out-innovating the rest of the world, and have catapulted the United States back to its leadership position in space, expanded our economy, and improved our national security. SpaceX has already transported our astronaut crews to the Space Station at a cost an order of magnitude lower than all previous human spaceflight missions. The government programs and policies that incentivized these successes were not preordained.

Garver and the pioneers she calls "space pirates" recognized that reducing the cost of space transportation was critical to fully utilizing the vantage of space for society. A confluence of the right people, technologies, policy, and private capital combined to pry a fraction of NASA's human spaceflight budget free from the powerful military-industrial complex to make these advances.

As often happens with scrappy entrepreneurial business competitors, Musk, Bezos, and Branson are prodding each other on. Competition ultimately benefits NASA and its big corporate contractor, the United Launch Alliance (ULA), a venture between Lockheed Martin and Boeing. The relationship is symbiotic. The greatest technological advances come from combining the resources of a visionary government with the scrappiness of risk-taking entrepreneurs.

Transformative change in government is often sought, but rarely achieved. In this revealing and personal book, Garver tells the fascinating story of how she helped a band of dreamers, rogue bureaucrats, and billionaires usher in a new space age.

—Walter Isaacson

TIMELINE OF KEY EVENTS IN HUMAN SPACEFLIGHT

1957	Russia launches first satellite to space, Sputnik
1958	President Eisenhower signs law establishing NASA
1961	Yuri Gagarin becomes first person to go to space and orbit Earth (April)
	Alan Shepard becomes first American to go to space (May)
	President Kennedy proposes sending a man to the Moon to Congress (May)
1962	John Glenn becomes first American to orbit Earth
1967	Apollo 1 fire kills three astronauts in training mission
1968	Apollo 8 first circumnavigates the Moon
1969	Apollo 11, Neil Armstrong and Buzz Aldrin become the first people to walk on the Moon
1971	President Nixon proposes the Space Shuttle program
1972	Apollo 17, Harrison Schmitt and Gene Cernan walk on the Moon in final Apollo mission
1973	Three astronaut crews launch to first US space station, called Skylab, staying in orbit for a combined 171 days
1975	US astronauts and Russian cosmonauts shake hands in orbit on the Apollo-Soyuz mission
1981	First Space Shuttle flight
1984	President Reagan proposes Space Station Freedom
1986	Space Shuttle Challenger explodes on liftoff, killing seven astronauts
1988	Space Shuttle returns to flight

1989 President Bush (41) proposes Space Exploration Initiative to the Moon and Mars

1993 Space Exploration Initiative terminated

President Clinton proposes adding the Russians as partners on the Space Station. The name later changes from Freedom to International Space Station (ISS)

1996 NASA awards Lockheed Martin a cooperative agreement for the X-33 program

2000 First launch of sustained astronaut crews to the International Space Station (ISS)

2001 X-33 program terminated

2003 Space Shuttle Columbia breaks apart during re-entry, killing seven astronauts

2004 President Bush (43) proposes retiring the Shuttle by 2010 and returning humans to the Moon by 2020 (January)

Paul Allen and Burt Rutan's SpaceShipOne wins the $10M Ansari X-Prize (October)

2005 Space Shuttle returns to flight

2006 NASA awards contracts for Constellation program

2009 American Recovery Act provides initial funding for NASA to start a commercial crew program

2010 President Obama proposes cancellation of Constellation and establishment of Commercial Crew program (February)

NASA awards first round of Commercial Crew agreements (February)

Legislation directs continuation of Constellation contracts, creating what becomes the Space Launch System (SLS) (October)

2011 NASA awards second round of Commercial Crew agreements (April)

Final Space Shuttle flight (July)

NASA provides Congress with plan for utilizing Constellation contracts to build SLS (September)

2012 NASA announces round three of Commercial Crew awards

2014 NASA awards Commercial Crew contracts to Boeing and SpaceX

2018 Virgin Galactic's SpaceShipTwo successfully travels to and from space with two pilots

2019 The Trump administration announces policy to return astronauts to the Moon by 2024

2020 SpaceX launches and returns the first successful Commercial Crew mission (May/October)

SpaceX begins regular Commercial Crew operations to ISS (November)

2021 NASA selects SpaceX to build its Human Landing system for the Moon (April)

Virgin Galactic carries its founder, Richard Branson to space with three other passengers and two pilots (July 11)

Blue Origin carries its founder, Jeff Bezos to space with three other passengers (July 20). They carry a total of ten more people to space on tourist flights in October and December.

Boeing experiences continued problems with its commercial crew Starliner, test flight remains grounded (August)

SpaceX carries out first entirely private citizen orbital spaceflight with four person Dragon flight (September)

PROLOGUE

THE MARS SCIENCE LABORATORY WAS ON TRACK TO LAND ON THE RED Planet at 10:30 p.m. PDT on August 5, 2012, after traveling over 10,000 miles per hour for 283 days. The mission—known as MSL—carried the largest, most complex and scientifically advanced spacecraft ever built to land anywhere other than Earth. The rover's name was Curiosity, and its charge was to determine if the conditions for life had ever existed on our neighboring planet, thereby shedding light on humanity's place in the universe.

The most critical phase of the mission was entry into the Martian atmosphere. After traveling through space at temperatures as low as 455 degrees Fahrenheit below zero, the spacecraft carrying the car-sized roving laboratory would be heated to 2,300 degrees, slowing her speed to a soft touchdown. Only about half of the missions ever sent to the Red Planet had successfully gone the distance and returned signals once they arrived.

The rover's size required a newly invented landing system of parachutes, a crane, and retro rockets to perform a precise choreographed automated routine that took seven minutes which NASA described as "seven minutes of terror." Even though the landing signal is transmitted to Earth at the speed of light, the very long distances meant that by the time Earthlings received the signal that Curiosity had touched the atmosphere, the rover would have already been alive or dead on the surface. I'd experienced both failed and successful Mars landings while working at NASA, but I felt a special sense of attachment to this mission.

The Jet Propulsion Laboratory (JPL) built and operated the spacecraft for NASA, and the head of the Lab, Dr. Charles Elachi, had briefed me on problems with MSL nearly four years earlier during the fall of 2008. I was then leading the transition team at NASA for the newly elected Obama administration, and the mission was scheduled to launch the following summer. MSL was already $400 million over its $1.5 billion budget, and the window of opportunity to send spacecraft to Mars opens for a few weeks every twenty six months. Missing the upcoming window would delay the launch until 2011 and increase the price tag to $2.5 billion dollars—a 60 percent cost overrun.

My transition role at NASA was advisory, so I was not the decision maker on the issue. But MSL would be carried out by the new administration taking over in a few months, so the JPL Director had come to Washington, DC, to get a read on the appetite the future President might have for a schedule slip. It wasn't my call, but Charles asked how I thought the incoming administration would want to proceed. My reaction was unequivocal. If it were up to me, the team should not be pushed. The best plan was to take the time and resources necessary to give it the utmost chance for success. In my view, landing a $2.5 billion dollar spacecraft successfully on Mars was infinitely better than losing a $2 billion dollar mission.

• • •

The extra time was taken, allowing the engineering and science teams to comb through every aspect of the spacecraft, and MSL was finally on the launchpad in November 2011. By then I was the NASA Deputy Administrator and on hand to escort VIPs and show my appreciation to the team. Curiosity lifted off on her journey to Mars from the Kennedy Space Center in Florida two days after Thanksgiving. With its successful launch complete, the next waiting game began.

Eight months and 350 million miles later, as the spacecraft closed in on her destination, all the action was on the other side of the country, in JPL's Space Flight Operations Facility (SFOF) in Southern California.

I'd arrived a day early to speak to well-wishers gathered at the Pasadena Convention Center for a conference called Planetfest. The event was held every few years, timed to coincide with extraordinary occurrences of cosmic magnitude. I'd attended my first Planetfest when Voyager—one of

NASA's longest running interplanetary robotic explorers—had its closest encounter with Neptune in 1989. Planetfest was sponsored by the Planetary Society, an advocacy and educational organization founded by the late Dr. Carl Sagan in 1980.

I'd been a member since the mid-1980s and had the chance to work with their leadership, including Carl Sagan himself before his untimely passing in 1996. Dr. Sagan believed that the discovery of life forms beyond planet Earth would be transformational; he was a free spirit who relished speaking truth to power. He passed on his values and beliefs to me and millions of others. I was thrilled when Bill Nye (the Science Guy) became president of the Planetary Society in 2010. He was already a good friend, and, as a former student of Carl's, it seemed fitting that he was at the helm. Bill had initially invited me to speak at Planetfest about NASA space policy. He followed up to ask if I would also help the Society honor Dr. Sally Ride, who passed away two weeks before the conference.

I knew it would be a challenge to speak about Sally without becoming emotional, but I gathered myself in order to honor her legacy. Sally played an important role in the space program and in my own life. I focused my talk on how she influenced the programs that were allowing NASA to innovate to make greater strides for the future of humanity. Like Carl Sagan, Sally cared more about having a positive influence on what was to come than reminiscing about the past.

Seeing several hands still raised with questions after my prepared remarks, I looked for my executive assistant Elise Nelson to let me know if I was running over my allotted time. I found her standing near the stage with a security guard, so I did my best to wrap up and find out what was happening.

When I approached Elise, the guard told me we needed to follow him and leave the room quickly. As usual, there were a handful of people waiting to talk with me after my speech, and though it felt rude not to linger, I did what I was asked and kept walking.

My mind went first to some sort of catastrophic event on the International Space Station (ISS), or with one of our Earth-based facilities. Then I thought about Curiosity. Had we lost the spacecraft after getting so close? When I inquired where we were going, the guard said I would be told as soon as I was safe. His comment made no sense. If I wasn't safe, there must be some sort of bomb threat, so why was I the only one being escorted out?

I wasn't even being taken outside but rather to another part of the building. We reached an empty conference room and the guard ushered us in, saying he'd be at the door if we needed anything. We were to stay inside the room.

Once Elise and I were alone, she said there had been a threat made against me and she had been told to get me out of a public space immediately. NASA security in Washington, DC, had directed her to have me call them when I was "secured." I was shaking when I placed the call, starting to realize that the situation might be serious.

The NASA security team explained that a threatening letter with a white powdery substance had been received at NASA Headquarters addressed to me. The person who opened the envelope in the mail room was being held in quarantine while the substance was tested. I was to remain in lockdown until the threat level could be assessed.

Elise and I had formed a close bond after sharing many wonderful experiences in our travels and we did our best to make light of the situation. The wait wasn't long before my phone rang and we were told the test had returned a negative result for anthrax or any other toxin. I was relieved for the mailroom personnel who'd been exposed and wondered what the letter said that caused security to think I needed protection 2,300 miles away.

I'd made some powerful enemies in my first three years on the job, but this was the first time I knew of physical threats. The NASA security team took the safety of the Agency's leadership seriously, and I hoped it was just an overzealous response to a random act. I turned my attention to the more immediate and interesting topic to contemplate; the spacecraft headed for its attempted Mars landing the following day.

On August 5, 2012, as I watched the JPL team from the viewing gallery over mission control, my thoughts were with the people who had spent over a decade of their lives working on the mission. Their success had the potential to reveal answers to some of humanity's oldest and deepest questions. As the Curiosity rover approached its atmospheric entry, I braced for the spacecraft's own seven minutes of terror.

PART ONE

GRAVITY

def. The universal force of attraction acting between all matter; the attraction of bodies toward the center of the Earth; great seriousness

1.

GAME CHANGER

THE FIRST SUBSTANTIVE CONVERSATION I HAD WITH BARACK OBAMA ABOUT NASA was in June of 2008, when he had just become the presumptive Democratic presidential nominee. I was introduced to him as the former space policy advisor to the Clinton campaign and the introduction seemed to pique the Senator's interest. He told me his "friend Ben Nelson had been lobbying him to extend the Space Shuttle" and he asked if I agreed with the recommendation. There were two Democrats named Nelson in the Senate at that time—Ben, from Nebraska and Bill, from Florida. I responded, "I think you mean Bill" and "no, I do not agree." I hadn't intended for the remark to come off as disrespectful, and when he shot me his big signature smile, I was relieved to see he had taken no offense.

Quickly acknowledging that it was indeed Bill Nelson who had been lobbying him, he asked me why I didn't think we should extend the Shuttle. I explained that while the Shuttle was the most visible part of NASA, its designated purpose—set over thirty-five years before—had been to lower launch costs and make space travel routine. Regrettably, it had never come close to achieving this goal. I reminded him of the loss of two astronaut crews and the accident investigation board's recommendation that it be retired in 2010. I noted that the Shuttle was built on forty-year-old technology. Although it was designed to fly 40 to 50 times a year, it had only flown an average of five in its first twenty-seven years, at a cost of over $100 billion dollars. He listened to my rant and then asked, "What do you think we should do instead?"

Now it was my turn to give him a big smile, as I walked him through how I thought NASA could drive advanced technologies and cutting-edge science to better fulfill its promise to the American people. Instead of competing with the private sector by doing the same thing over and over, I suggested that incentivizing companies to take on the routine aspects of the program would free NASA to invest in programs of greater relevance to the taxpayer. I explained that NASA was formed to utilize the vantage of air and space to benefit the public, yet its programs to address our most current problems—such as those related to climate change—received less than ten percent of its budget. Allowing companies to open new markets would not just lower costs for more consequential research activities in space; the policy shift would produce broad economic and national security gains. If it was an interview, I knew I passed when I got a call a few weeks later asking me to lead the NASA transition team if he became President-elect in November.

I'd spent my twenty-five-year career training to be prepared for such an assignment, and although my background was different from everyone who had been in the position before, I believed that was a positive feature and not a bug.

I hadn't been drawn to a career that involved space in order to build rockets or become an astronaut. I was attracted by the unlimited potential space activities offered our civilization. I was a child of the 1960s who loved a challenge, and by the early 1980s, when I was just starting out, space seemed like the most meaningful challenge ahead. After running the gauntlet of deterrence by high school teachers and counselors against entering male-dominated science and engineering fields, I pursued degrees in political economy and international science & technology policy. Determined to make a difference, I saw space as a blank canvas full of value and endless opportunity.

The rare alignment of the planets that allowed me—someone with a less traditional background—to lead President Obama's NASA transition team, came at a significant point in history. Lured by the prospect of a growing space economy and frustrated by the lack of government progress, daring individuals were developing innovative technological advancements in spacecraft and space transportation that were beginning to succeed. I thought NASA should build bridges to these new entrants and ideas that could finally make space more accessible. Being assigned to this position gave me the opportunity—and the obligation—to ensure that the administration put forward policies and programs that would shift the paradigm and usher in greater progress.

I recruited a small volunteer team, and we began gathering information on current NASA activities, highlighting strengths and weaknesses of alternative paths, while teeing up options for more meaningful programs. Our final transition report was consistent with my initial conversations with the President-elect and closely aligned with his cross-government policy focus on science and innovation. It offered a transformative agenda that would reduce the barriers to access space and allow the public to reap the benefits of their investment.

Our report was so well received by the incoming administration that soon after his inauguration, the President expressed his intent to nominate me for NASA Deputy Administrator. He selected Steve Isakowitz as his intended nominee for the NASA Administrator position a few weeks later. Steve had topped my recommended list of people to lead the Agency, and his selection was affirmation of the administration's alignment on its vision for NASA. Steve Isakowitz had multiple technical aerospace degrees from MIT and twenty years of experience working in the aerospace industry. He'd held senior positions at NASA, the Office of Management and Budget, the CIA, and the Department of Energy. Steve had served in both Republican and Democratic administrations and was widely respected by the community. His qualifications for the position were undeniably impeccable.

The White House planned to put our nominations forward simultaneously. Vetting procedures got underway, and we began to discuss how to develop a bold, sustainable plan. I hadn't been an early supporter of candidate Obama, but I was already seeing how reshaping space activities could help translate his campaign's "hope and change" mantra into more than a slogan. The Space Age envisioned fifty years earlier finally seemed within our grasp. All presidents dream of being transformational, and in February of 2009, I believed NASA could make that dream a reality for the Obama administration.

The first disturbance in the force came when Senator Bill Nelson declined to schedule a meeting with us. The Florida Democrat's stated reasons were nebulous and didn't involve me. The White House personnel office relayed to us later that the Senator had his own candidate. I didn't consider the threat seriously at first, believing the President's clout was sufficient to withstand foot dragging from a single Senator within his own party—especially for someone with Steve's qualifications.

The Democrats controlled the Senate with 60 votes, so confirmation of virtually any NASA nominee was a near certainty. Nelson wasn't even the committee chair responsible for holding the hearings. That was Senator Jay Rockefeller, a conservative Democrat from West Virginia. Rockefeller was a rare congressional overseer of the space agency—he had an open mind. He would clearly have had an open hearing docket for any NASA leadership team the new President put forward.

The White House could have proceeded without Senator Nelson's support and scheduled our pre-confirmation meetings with Senator Rockefeller and other members of the committee, but these were the early days, when they hadn't yet learned they'd need to fight for every ounce of progress. The personnel team told Steve they would consider a temporary appointment that would likely lead to later confirmation, but without the President's willingness to take on Senator Nelson directly, Steve stepped aside.

I couldn't believe a single Democratic senator's personal views were enough to sideline the President's extremely well-qualified nominee. It didn't bode well for progress.

Bill Nelson was a lifetime politician most known for his out-of-this-world political junket in 1986: a taxpayer-funded ride on the Space Shuttle. Like other members of Congress from Southern states with NASA facilities in their districts, his interests often appeared parochial. When I'd been told by candidate Obama the year before that Nelson was lobbying him to extend the Space Shuttle program, it appeared to me that his agenda was shortsighted.

An investigative review board of the 2003 Space Shuttle Columbia accident recommended retiring the Shuttle fleet by the end of the decade, and President Bush had agreed, establishing the policy in 2004. I supported flying one or two more missions, but fully reversing that decision in 2009 would have taken several years to implement, cost billions of dollars, and risked more astronauts' lives. The NASA briefings I'd received during the 2008 transition period concluded that that ship had sailed. Worse, we'd learned that the planned replacement program—called Constellation—was badly off course. The new program was already costing $3–4 billion a year and had slipped five years in its first four years of development.

Constellation was established to support a long-term goal of returning a handful of astronauts to the Moon—something NASA had been hoping to do since the 1980s. It required an Apollo-sized budget but lacked a geopolitical or other rational, national purpose. Instead of driving technology as Apollo had, it was based on existing technologies—a reorganization of Space Shuttle parts and contractors.

Planned lunar missions were more than a decade away, so Constellation's stated initial purpose was to transport astronauts to and from the Space Station. Unfortunately, the rocket- and capsule-funding needs already exceeded any realistic budget. NASA's five-year plan put forward by the Bush administration and briefed to the transition team, was to make up the funding shortfall using money budgeted for the Space Station itself.

Defunding and therefore de-orbiting the Space Station early would leave the rocket and capsule without a destination. By the time the first elements of Constellation were ready to fly, the Space Station would have been charred fragments strewn across the bottom of the Pacific Ocean. Not only would NASA lose its ability to launch astronauts for many years, all of NASA's and its international partner's spaceflight activities would have ceased.

NASA's unstated plan was essentially to trap the next President into adding several billion dollars a year to keep money flowing to Shuttle, Constellation, and Space Station contractors. The human spaceflight side of NASA typically took precedence, so they also figured they could siphon more funds from Earth and space science to cover their overrun. Even then, no amount of money would be able to close the space transportation gap befalling human spaceflight. NASA's intention—known well to Congress—was to pay the Russian Space Agency—Roscosmos—to carry its astronauts to and from the Space Station after the Shuttle retired.

Human spaceflight was in an untenable situation and without new leadership arriving soon, precious time to map a more realistic course was slipping away.

One other senator weighed in on selection criteria for the NASA Administrator position early in the process, Senator Barbara Mikulski. Senator Mikulski (D-MD) was in many ways more important to NASA than Senator Nelson, since she chaired the Agency's appropriations subcommittee. At our first face-to-face meeting during the transition period, Senator Mikulski told me to relay the following to the President-elect: "No astronauts and no military people." It made sense to me, and I took the note. We discussed other topics, and before I left, she circled back to her comment on Administrator qualifications. She said, "No astronauts unless it's Sally Ride." When I relayed her sentiment to the personnel team, they asked me to see if Sally would be interested.

I'd gotten to know Dr. Ride through her extensive post-astronaut service to NASA and was under no illusions that she'd evolved her position since President Clinton tried to recruit her for the job eight years before. Our conversation went as expected. Sally knew the game and didn't want to play. She expressed her willingness to help in any other way, but practically begged me not to have Obama call her directly, since he'd be a lot harder to say no to than I was. I thought Sally would make a fantastic Administrator and knew that if she said yes, Senator Nelson would have likely supported her alongside Senator Mikulski. But Sally didn't want the job and we were back at square one. The White House continued to interview potential candidates for the Administrator position, but none made it very far through the vetting process, so the standoff continued.

The delay stalled progress at a crucial time in the budget process. Anticipating my own nomination as deputy, I had left my formal transition team role on January 20. I'd been able to oversee the development of NASA's portion of the stimulus bill, which included significant funding for our new priorities, but after I left, the acting Administrator worked with the Hill to transfer much of what was allocated to Constellation. Budgets for the following year had to be developed that spring, and without NASA's willingness to craft a more sustainable plan for human spaceflight, the administration needed a workaround.

In lieu of a new leadership team, we established a presidential committee to review the human spaceflight program and form a more realistic path forward. The administration appointed ten esteemed technical experts and policy leaders—including Sally Ride—to a group that became known as the Second Augustine Committee, named for its Chair, Norm Augustine, the former CEO of aerospace giant Lockheed Martin.

The human spaceflight review board was made public in May. A few weeks later, the President announced his nominee to run NASA, Charlie Bolden. Charlie was a marine general and astronaut who'd flown on the Shuttle with Congressman Bill Nelson twenty-five years earlier. My nomination for Deputy Administrator was concurrent, but unnoteworthy by comparison. We sailed through the process and were confirmed by Senate acclimation in July.

The Augustine Committee's findings were released a few months after we were confirmed. The panel found that "the US human spaceflight program appears to be on an unsustainable trajectory." Their report said NASA was "perpetuating the perilous practice of pursuing goals that do not match allocated resources." They outlined potential options that pursued new technologies to utilize the burgeoning commercial space sector to generate new capabilities and potentially lower costs.

The Augustine Committee's views—consistent with those of the transition team report—combined to inform and underpin President Obama's proposal to shift NASA away from developing and owning systems for routine operations and incentivize the private sector to provide space transportation services for cargo and astronauts—crew—allowing NASA to invest in more cutting-edge technologies and breakthrough scientific discoveries.

On February 1, 2010, the administration's first full budget publicly requested $19 billion for NASA to fly out the Shuttle safely and extend the Space Station; increase funding for Earth sciences, advanced technology, rocket engine development, and infrastructure revitalization; and begin a partnership with US industry to transport astronauts to the Space Station, referred to as Commercial Crew. The transformational agenda was structured to allow the Agency to begin to shed the institutional burdens that constrained progress, which required terminating its beleaguered Constellation program.

The established space supporters in Congress and industry were outraged by the plan. Entrenched aerospace interests had spent their careers designing versions of Constellation-like programs to keep expensive infrastructure and jobs in key congressional districts at the expense of more competitive programs, regardless of operational effectiveness. The companies with contracts worth tens of billions of dollars cried foul and combined their lobbying might against the plan. Ignoring numerous government audits and the public results of the Augustine Committee, traditional stakeholders argued we'd proposed radical changes that would damage the NASA institution. They claimed to be blindsided by the proposal.

The Administrator had difficulty explaining the proposal's merit, so it was assumed he hadn't devised or supported the strategy.

I became the target of the campaign against the plan.

I was attacked by Democrats and Republicans in Congress, by the aerospace industry, and by hero astronauts for proposing an agenda that didn't suit their parochial interests. The elation and promise of the administration's potential to drive meaningful change was already being threatened by the trillion-dollar military-industrial complex, and I was the one taking fire.

Senator David Vitter from Louisiana accused me of orchestrating the cancellation of Constellation, and suggested I "was running the Agency, and not the Administrator." Homer Hickam, author of *October Sky* and the subject of the motion picture *Rocket Boys*, called me a "gadfly who should resign." Senator Richard Shelby, the senior Republican on the appropriations subcommittee handling NASA funding, said that the President's proposed NASA budget "begins the death march for the future of US human spaceflight" and that "Congress cannot and will not sit back and watch the reckless abandonment of sound principles, a proven track record, a steady path to success, and the destruction of our human spaceflight program." In reference to the budget request for Commercial Crew, he said, "Today the commercial providers that NASA has contracted with cannot even carry the trash back from the Space Station much less carry humans to and from space safely."

As Chair of the Senate subcommittee that authorized NASA, Senator Nelson criticized the President for slashing the Moon program and said the move could cause the United States to fall behind other countries in space exploration—most notably Russia and China. He highlighted several positives in the budget request, such as extending the Space Station, but said the budget was not well received because it gave the perception of killing the manned space program for the United States. He admonished the administration for a lack of leadership and suggested the President had somehow allowed budget examiners to dictate his NASA agenda.

During a March subcommittee hearing on US commercial space capabilities, Senator Nelson asked repeatedly about the $6 billion intended to fund Commercial Crew taxis proposed in the budget request, asking "what would happen if Congress decided—since the Congress controls the purse strings—that we wanted to take the $6 billion projected by the President over the next five years and use that not for human certification of the commercial vehicles but instead to accelerate the heavy-lift vehicle for the Mars program?" The Florida senator was not alone in his opposition to our proposal, but he was NASA's most attentive and influential Democrat in Congress, and the President had already acquiesced to his demands on Agency leadership. Instead of acknowledging the value of the proposal and advocating for its substance, he coordinated opposition to it with his Republican counterparts.

Senator Kay Bailey Hutchison, a Texas Republican who held leadership positions on NASA's most important committees, opened an early hearing by saying, "Congress must examine closely the very underpinnings of the proposed NASA budget request, which I believe, if accepted and supported by the Congress in its present form, would spell the end of our nation's leadership in space exploration. That would certainly be the case in the area of human spaceflight capability."

She said she was "skeptical and very disappointed" in the proposal to provide funding to assist the development of a commercial launch vehicle, saying, "The emphasis to the tune of $6 billion into a very fledgling commercial capability I just think is not sound and it's certainly not going to be reliable." None of the program's critics admitted that the alternative plan they'd already approved was to send hundreds of millions of dollars to Roscosmos for the astronauts' rides to the Space Station.

Two of our greatest living American heroes—the first and last astronauts on the Moon, Neil Armstrong and Gene Cernan—testified in Congress that "the administration's budget for human space exploration has no focus and in fact is a blueprint for a mission to nowhere." Neil Armstrong added that the plan was likely contrived by a small group in secret and charged the President had been "poorly advised." Cernan said the proposal was most likely created in haste, with little input from the Administrator, by people promoting their own agenda. He charged that "not only is human spaceflight and space exploration at risk, but the future of this country and our children and our grandchildren as well." He concluded his testimony saying, "Now is the time to overrule this administration's pledge to mediocrity, now is the time to be bold, innovative and wise as to how we invest in the future of America."

I'd known Gene and Neil for two decades by this time, and like everyone else on the planet, I was in awe of their heroism and achievement. Like many early astronauts, they would have preferred the government had continued to spend Apollo-scale budgets to send a few astronauts like them farther into space continuously. Gene had coordinated the testimony and was a vocal Republican and critic of President Obama. They claimed their views weren't partisan or personal, but it was hard not to take it personally when Neil Armstrong wrote that "the transition team should play no part in such decisions. While these men and women are experienced and enthusiastic space program veterans, they are neither aerospace engineers nor former program managers and cannot be sufficiently knowledgeable to make choices in the technical arena."

I wholeheartedly agreed that space enthusiasts like me shouldn't be making technical decisions, but that was not my role on the transition team nor as NASA Deputy. My recommendations to the President had utilized credible, independent technical analysis to inform my policy and management advice, and aligned with the goals of the elected leadership. I hadn't been to space, but I'd studied the proper role of government and how to design policies to incentivize behaviors that were in the public's best interest—something not in the curriculum for astronaut training.

The epic battle that ensued pitted traditional space loyalists against a new generation of space advocates who believed NASA had been hijacked and needed rescuing. On one side were the large stakeholders—aerospace companies, lobbyists, astronauts, trade associations, self-interested congressional delegations, and most of NASA. On the other side—a handful of outspoken space enthusiasts and bureaucrats, a few billionaires, political appointees, and the President of the United States.

The movie *Moneyball* was released just as the criticisms and threats were at their peak. Hearing John Henry, the owner of the Red Sox, say to Oakland A's General Manager Billy Beane—who is played by Brad Pitt—that the first one through the wall always gets bloody, helped me accept and even embrace my scars. I've watched the scene a dozen times and it always gives me solace:

> I know you are taking it in the teeth, but the first guy through the wall... he always gets bloody... always. This is threatening not just a way of doing business... but in their minds, it's threatening the game. Really what it's threatening is their livelihood, their jobs. It's threatening the way they do things... and every time that happens, whether it's the government, a way of doing business, whatever, the people who are holding the reins—they have their hands on the switch—they go batshit crazy.

The transformation we were advancing was doing exactly that; it was threatening a way of doing business—a business worth hundreds of billions of dollars. Their protectionist response was to lash out and cast blame. Without visible support from the President or NASA Administrator, I became their target and was accused of threatening to destroy human spaceflight forever. The people holding the reins, in the institutions controlling the space program since the 1960s, went batshit crazy.

My quest to make space more accessible and sustainable wasn't meant to start a war. I wasn't trying to steal the future. I was on a rescue mission. It isn't just Earth's gravity that we must overcome, it is the gravity of our situation.

• • •

Formed by law in 1958, NASA launched the first American into space and was given the political mandate to send a man to the Moon in its first three years of existence. The young space agency successfully completed the Mercury, Gemini, and Apollo programs within the next decade. The Mercury program carried out twenty uncrewed and chimpanzee test flights and six successful astronaut flights; Project Gemini launched ten successful missions and sixteen astronauts into space; the Apollo program launched eleven crewed missions and twenty-nine astronauts while landing twelve men on the Moon and returning them all safely to Earth.

I was ten years old at the height of Apollo, when studies on the future of NASA predicted average citizens traveling to space affordably, colonies on the Moon, and placing people (men) on Mars by 1980. Nearly thirty years later, as I began to advise the President-elect on how best to reshape NASA policies, the United States had flown fewer than 350 astronauts to space and none had traveled farther than 400 miles overhead. The average cost to the taxpayer for each astronaut flown was over one billion dollars, and the cost to develop and launch robotic spacecraft and satellites remained similarly astronomical.

In inflation-adjusted relative terms, NASA's budget has always been more than half of its peak when it was part of the Cold War effort to beat the Soviets to the Moon, yet the space community blamed insufficient funding for its lack of progress. NASA's leaders were typically astronauts and engineers who didn't question the public value or relevance of their activities. Indeed, many considered flying themselves and their friends into space to be an entitlement that shouldn't require justification. They had little interest in transitioning what they enjoyed and got paid to do over to the private sector and they assumed that was their decision.

For NASA to reach its full potential, I believed we needed to realign government programs and policies to make space activities more sustainable. This required reducing infrastructure and transportation costs, assuring the long-term ability to access space, and safeguarding the environment to help keep Earth (and space) habitable. To others, sustainability meant spreading work around to key congressional districts to ensure existing programs could not be canceled. My varied experiences gave me an outsider's perspective that I knew was hard for the aerospace community to embrace, but I also knew that the larger cause was worth fighting for.

I grew up enchanted by NASA's history, but I saw that by putting our past achievements on a pedestal, we'd limited our future. We were oriented toward challenges that had long since passed. NASA's overwhelming early success had narrowed its field of vision and prevented the space program from advancing at its predicted pace. The space community was frustrated by the decades of stagnation but didn't want to accept the fundamental source of the problem. Too many stakeholders were invested and incentivized to protect the current way of doing business.

In reality, there are as many motivations to advance space development as there are people on the planet. That is the point. Instead of benefiting a few, space should be fully utilized to benefit humanity and society. Future space activities can help us not only to thrive but also to survive. The direct global market from activities that operate in the area we call space—beyond Earth's atmosphere—already returns nearly a half a trillion dollars in economic value and you don't have to major in math or science to understand that being limited to a single home world makes humanity much more vulnerable than a multi-planet species.

Space is an inimitable location with unique characteristics—just like Earth's own atmosphere and oceans. Industries that developed around exploiting the oceans and the atmosphere support vastly different but important uses, including transportation, communications, scientific research, national security, tourism, and recreation. Early explorers pioneered an expanded ability to operate in these otherwise hostile, alien environments.

Industries geared toward exploiting space are evolving to support similarly diverse and important uses. Operating in space provides us instantaneous global delivery of voice, data, and video information; precision measurement of time and location; and Earth observations that measure the interactions between the atmosphere, land, ice, and oceans that affect us all. Space endeavors contribute to the greater good by connecting the world and reaching beyond, improving our knowledge, economy, and national security. NASA's early successes in human spaceflight enhanced US prestige, solidified our global leadership position, and provided inspiration to people around the world.

Different traversable natural environments require their own transportation systems to navigate safely. As traveling in space becomes more reliable and cost-effective, a multitude of vehicles, uses, and destinations will be created to better capitalize on its perspective, conditions, and resources, just as has happened in shipping and aviation.

Unlike in shipping and aviation, the government remained the dominant force in human space transportation for more than fifty years. The systemic issues that limited progress at NASA and other government agencies are pervasive. Those in power have long believed that escaping gravity should remain in their control.

The truth is that advancing technologies and reducing barriers to entry for the US private sector, thereby allowing them to better compete internationally, are crucial roles of the government. As industries mature, government regulations related to public safety, environmental stewardship, and shared resource allocation must evolve to keep up with the pace of new capabilities.

Important technological achievements have been accomplished by individuals and private interests throughout US history. The Wright Brothers, Glenn Curtiss, Howard Hughes, Bell Labs, Steve Jobs, and Bill Gates—through innovation and investment—transformed technologies beyond exclusively government efforts, greatly contributing to society and our national interests.

The progression of human spaceflight has been impeded by competing government programs and improperly structured government policies that disincentivized the level of capital investment needed for its development—that is, until recently.

Space activities are finally on the cusp of achieving what aviation achieved in its first few decades. Those who can afford it are going to space just for the view and thrill of weightlessness—modern-day barnstormers. Stimulating what private companies like SpaceX, Blue Origin, Virgin Galactic, and hundreds of others are achieving today paves the way for more worthy advances to follow, but it is only a first step.

We now have the knowledge, understanding, and capability to chart a course that fully utilizes the realm of space to help manage Earth's resources sustainably. If we succeed, the most adventurous among us will one day join robotic explorers and expand humanity outward.

• • •

As a girl growing up in Michigan in the '60s—the daughter of a homemaker and stockbroker, and the granddaughter of farmers—a controversial NASA career wouldn't have been predicted. My memories of watching the Moon landings on television are dim. My mom kept a drawing I made of an astronaut standing on the Moon holding the flag next to the lunar module, but I would never have imagined how well I'd get to know the astronaut I had drawn. I have tried to conjure some deep-seated connection to space in my eight-year-old self, but I mostly remember playing with Barbies. If my parents had worked to foster my early childhood interests into a career, I would have likely become a beautician—foreshadowed by cutting off my doll's hair.

Without brothers in those days, you weren't likely to have airplanes, rockets, or space toys around the house, and we didn't have brothers. The closest I came to my yet-to-be-chosen field was a fleeting interest I had in the 1970s to become a stewardess, inspired by an uncle who was a United Airlines pilot. When Uncle Bruce showed my fifth-grade class the cockpit of his Boeing 737 during a field trip to the Lansing airport, the boys in the class received wing pins that read "Future Pilot" and the wing pins given to the girls read "Junior Stewardess."

When I was twelve, members of my church confirmation class privately selected a word that best described each other, to be revealed while we were confirmed in front of the congregation. We selected words like *leader*, *intelligent*, and *sportsman* for boys, and *cheerful*, *graceful*, and *nice* for girls. I stood in front of the congregation, anxious for the minister to announce the quality my friends saw in me. I had to fight back tears when I heard the minister say it—*determined*. While my determination served me well later in life, as a twelve-year-old girl in 1973, I wanted to be known as cheerful, graceful, or nice. As it turns out, my friends knew me better than I knew myself.

Looking back, I could have been called a tomboy; a generally negative label given to girls who exhibited characteristics and behaviors considered typical of a boy. I wore my hair short, loved sports of all types, and was good at them. I loved science and math, and I was good at them, too. But I also enjoyed doing ballet, playing music, and being a cheerleader. I wanted it all, and in the 1970s in mid-Michigan, to me that seemed possible.

I was a straight-A student in high school, and aptitude tests predicted engineering and science careers for me. There were six of us in my grade who had completed all the math available before our senior year of high school. When we returned from summer break, I discovered that the school administrators had registered the other five—all boys—in a calculus class at the local university. I hadn't been contacted about taking the course with them, and when my parents asked why, they were told it hadn't occurred to them that a girl would want to take calculus. My mom was particularly upset about this, but I was just as happy to add another elective to my schedule and was relieved I wouldn't have to commute to take the class with the geeky boys. But not taking calculus in high school channeled me into the social sciences in college, and like many girls my age, I didn't take much of a direct interest in space until NASA sent an astronaut there who looked like me.

That was the year, 1983, when I graduated from Colorado College and started my first full-time job working on John Glenn's presidential campaign. It is often assumed that I came to Washington, DC, to work for John Glenn because he was an astronaut. I haven't always corrected that assumption as I am aware that it fits into a nice narrative for my own mythology. The reality of my first post-college job was more pragmatic. I was disillusioned with the current national political leadership and wanted to help someone get elected who I thought would be better. More than a year before the election, as I was making post-graduation plans, John Glenn was the only Democrat in the field running ahead of President Reagan in the polls.

Not only is politics in my blood, but I've been campaigning since before I could walk. In addition to farming, my grandfather and uncle, both Republicans, had been in the Michigan state legislature for a combined forty years. My sister and I were featured on campaign brochures, and when I was a baby, my grandpa carried me while shaking hands in local parades. When the state legislature was in session, my grandpa let me join him on the House floor during school visits to the Capitol, and the experience left an indelible positive memory. My formative role models were public servants dedicated to helping their neighbors. Doing something similar became my aspirational goal.

By '83, I'd worked on a lot of political campaigns but never anything like a national election, and I found the experience invigorating. I eventually worked my way up to become a scheduler and spent long hours in a bullpen of desks adorned with overflowing ashtrays and large, constantly ringing phones with long, spiral cords attached to the receivers. I fell in love with the campaign and with my soon-to-be husband Dave Brandt, a recent Kent State graduate who worked in Glenn's press office.

John Glenn was the first astronaut to leave NASA, less than two years after his solo flight, and avoided serving on space-related committees as a senator. He wanted to be known for more than his three-orbit, five-hour spaceflight. But when *The Right Stuff*, the film based on Tom Wolfe's book—was released in theaters during the campaign, he agreed to exploit his spaceflight as a differentiator. It didn't pan out as he planned. Glenn was portrayed as a do-good outsider among the other Mercury astronauts, and some thought the movie did more harm than good.

When Super Tuesday results came in from thirteen states in March of 1984, after outspending the others, Senator Glenn didn't win a single state. A political cartoon ran the following day that caricatured Gary Hart saying, "I'm New," Walter Mondale saying, "I'm Ready," and John Glenn saying, "I'm History."

I hadn't gotten to know the Senator extremely well, but he was a politician, so he always pretended to remember me when he called or visited the office. My career led me to work with him several more times, both at NASA and advising other presidential candidates. My early campaign association with him provided a positive foundation for our continued professional relationship, even though his policy views were more traditional than my own.

With the campaign abruptly ending, senior staff looked out for more junior employees and helped me get an entry-level job at a nonprofit membership organization called the National Space Institute. Wernher von Braun, known as the father of the Moon program, founded the association with aerospace industry funding in 1974, frustrated by the lack of public and political support for NASA after Apollo. Von Braun had died in 1977, but my new boss, Executive Director Dr. Glen Wilson, had known him since his career began as a legislative clerk for Senator Lyndon Johnson. Dr. Wilson planned to retire soon and spent his mornings reading the newspaper and his afternoons sharing stories about his memories of von Braun and the early days of the space program, including what led to NASA's formation.

Before Dr. Wilson's retirement, the National Space Institute merged with another space advocacy organization, the L5 Society, and changed its name to the National Space Society. In stark contrast to NSI's top-down, industry-supported beginnings, the L5 Society was founded by a group of followers of Gerard O'Neill, a physics professor at Princeton University who had, among other things, developed a concept of free-floating, self-sustaining space colonies. The society's name came from Lagrangian points in the Earth-Moon system proposed as locations for the huge rotating space habitats that O'Neill envisioned.

NSI had focused on advocating for whatever programs NASA put forward, but the L5ers were activists who wanted to move the program toward more sustainable space development. The shorthand version is that von Braunians are explorers, drawn to space activities for the daring challenge, and O'Neillians are exploiters, drawn to space for economic expansion and human settlement. In addition to differences in what they supported, they differed on how they went about doing it. NSI had a top-down traditional approach, while the L5ers were activists willing to challenge the status quo. This was my first introduction to the group I refer to here as space pirates.

Similar to pirates on the high seas, space pirates have been depicted as both heroes and villains. The recurring villains in the 1930s Buck Rogers comic strip were called space pirates, but early science fiction authors used the term to refer to heroes mining asteroids and collecting other bounty on space trade routes of the future. More familiar references include *Star Wars*' Han Solo as well as Mark Watney, the character in the book and subsequent film *The Martian* by Andy Weir, who refers to himself as the first space pirate as he takes possession of a spaceship parked on Mars in "international waters," without permission of NASA, to survive on Mars. In 2019, after Senator Ted Cruz justified the Trump administration's new Space Force by claiming that just as pirates threaten the open seas, the same thing is possible in space, Elon Musk tweeted a picture of the pirate's signature flag with skull and crossbones.

As with any group, the space pirates are unique individuals who share some common characteristics and views. Many of them have spent decades working to create a spacefaring civilization at great personal cost. They have advanced important policies and legislation, kept the United States from signing treaties that would have blocked space development, started new companies and organizations, lobbied members of Congress, antagonized senior aerospace industry leaders, and often been ignored and marginalized by the established space community. These are the people who raised me—my original space family.

When the Space Shuttle program was announced in 1972, President Nixon said it would be "an entirely new type of space transportation system designed to help transform the space frontier of the 1970s into familiar territory, easily accessible for human endeavor in the 1980s and '90s." He said, "It will revolutionize transportation into near space, by routinizing it. It will take the astronomical costs out of astronautics." NASA's initial estimated $6 billion development cost quadrupled and by the mid-1980s it was obvious to anyone paying attention that it was never going to deliver on its stated promise.

The space pirates saw early that the biggest obstacle to space development was the lack of affordable, reliable access to space. They believed the Space Shuttle was impeding progress. To some this made the space pirates heroes and to others villains. One of the initial ways they advanced their objective was to devise the 1984 Commercial Space Launch Incentives Act, which was enacted to support acquiring more innovative equipment and services offered by the private sector. I viewed this as an extremely logical concept, without questioning why it was left to a small nonprofit advocacy organization to champion instead of the nation's space agency. I didn't yet realize what the space pirates already knew, that the traditional aerospace community cared more about increasing the size of their slice of NASA's budget pie than lowering costs to create a sustainable program.

Like most others, I still held NASA on a pedestal. My National Space Society office was across the street from NASA's DC Headquarters. My colleagues and I frequented the local bar, where we got to know astronauts and Agency leaders. I immersed myself in all things NASA and even played on their softball team. The Shuttle program succeeded in exciting the public initially, and as I got to know astronauts who weren't all white men from the military, I sensed we were nearing a new space age. NSS developed membership tours and public education activities focused on the popular new space plane, and I jumped at the chance to be involved.

I was leading a tour in Florida when Columbia launched mission 61-C on January 12, 1986. This was the mission's fifth launch attempt, and I had been corralling tour groups for all four delays, starting the previous December. Technical and weather issues had been plaguing the program, and after five years, this was only its 24th mission. NASA was anxious to prove they could increase the launch rate as advertised, and in an attempt to demonstrate the system was safe and routine, had begun flying non-professional astronauts on the Shuttle. As part of that effort, Florida Congressman Bill Nelson was a member of the crew that morning, along with rookie pilot Charlie Bolden. Never in my wildest imagination could I have envisioned how the bond they developed on that flight would impact the space program and my own career.

As an indicator of NASA's intent to pick up the pace, a second shuttle—the Challenger—was already sitting on the adjacent launchpad. As Columbia finally lit the candle, the next mission was scheduled to launch two weeks later.

I was thrilled to be back in the Florida sunshine leading another launch tour for the Challenger mission, set to take off on January 27. I waited with my tour group at the viewing site four miles away, answering their questions about NASA and the Shuttle program as morning turned into afternoon. The technicians closing the hatch had trouble removing the handle from the vehicle's door, and after trying to unstick the handle manually for a while, they requested power tools to help with the removal. Pad technicians went to retrieve battery-operated drills and cutting blades, only to find them drained of power once they got to the top of the gantry. Next, they decided to cut the handle with a hacksaw, and they called for another delivery. Again, a maintenance worker ran from the service building and took the elevator up to the gantry where the team finally just sawed off the handle. By now, a weather front had blown in and the winds exceeded the launch threshold. The hours-long comedy of errors left no time for the storm to blow through, so the launch window expired, and the seven astronauts were escorted out of the vehicle.

Increasing public interest and proving the Shuttle was safe and routine wasn't just about flying members of Congress; NASA started a program to fly average citizens, beginning with the Teacher in Space. The first teacher—Christa McAuliffe—was a member of the Challenger crew, which made the day's troubles even more publicly embarrassing. I had met Christa and several other crew members at events in Washington and imagined they were as frustrated as anyone by the confounding delay.

As we were leaving the viewing site, I asked the NASA volunteer assigned to our bus whether or not he thought they would try to launch the next day. The twenty-something-engineer casually responded that the weather forecast indicated it would be too cold for a launch the following morning. With that information, I took a late flight back to DC. I was in my apartment getting ready to head into the office when I saw the countdown had begun for another launch attempt.

A cold front had indeed rolled into Cape Canaveral overnight and cameras were showing photos of ice on the vehicle and tank throughout the fuel's loading. It didn't appear to be a problem, since they proceeded with the countdown. I was disappointed to miss what was always a thrilling experience in person and a bit irritated at being told bad information about a potential weather problem. Disappointment turned to disbelief when the Shuttle's contrail exploded into a fireball 73 seconds into the flight.

The Challenger broke apart as the astronauts' families and loved ones attending the launch searched the empty Florida sky and millions of school children watched on television. NASA's first in-flight astronaut fatalities were on a vehicle that had promised cheap, routine space transportation and was deemed to be safe and reliable.

Later, when Sally Ride and others zeroed in on the temperature being the problem that caused the hot gasses to escape from the solid rocket motor, the world learned that NASA managers and contractors who were in charge had overridden vehement objections from engineering and waived the established temperature rules. Like many others, I was astonished and disheartened that NASA and industry leaders had been so reckless with the nation's precious assets—the lives of the crew and the future of human spaceflight.

The Challenger accident was a determinative event for space development. In order to justify the government's large investment in the program, US policy had directed nearly all its satellites to be launched by the Shuttle, which had extinguished the competition. The disaster not only killed seven astronauts, it grounded scores of national security, civil and commercial satellites. The accident led to a new policy that directed the Shuttle be used exclusively for missions that required the presence of astronauts and the government started transitioning ownership of the existing expendable rockets to the private sector. After nearly a three-year hiatus, the Shuttle returned to flight with a more insular mission.

President Reagan had initiated a program called Space Station Freedom in 1984, designed to be NASA's central purpose for the Space Shuttle. It was billed as "The Next Logical Step." Without a destination, the Shuttle limited human spaceflight to week-long missions, so developing a station was critical to learning how to live and work in space for longer periods of time. Not coincidentally, having a space station helped justify continued operations of the Shuttle. There wasn't much of a debate about whether to continue with the Shuttle after Challenger, but without a space station program on the drawing board, that might have been a different story.

The Space Station's goals—outlined by President Reagan's introductory speech for the program—were for scientific advancement and commerce. Reagan predicted the "space station will permit quantum leaps in our research in science, communications, and in metals and lifesaving medicines which could be manufactured only in space." He said, "Just as the oceans opened up a new world for clipper ships and Yankee traders, space holds enormous potential for commerce today." By the time I returned to work at NASA twenty-five years and more than $100 billion later, achieving these goals had remained elusive.

The 1986 Shuttle accident was also determinative in my own career. Like it did with others, the tragedy caused me to question what NASA was doing and to what end. NSS was one of the few non-government space organizations in those days, and we were called upon to provide expertise and to field media requests. I was immediately thrown into the deep end of the pool, appearing as a guest on DC's local NPR station the evening of the accident. When I didn't drown, I received more requests to serve as a public space analyst and spokesperson.

I enjoyed highlighting the many innovative industries and unique scientific information that had been gained from our space endeavors. There was a general appreciation for how NASA's early investments helped drive instantaneous global communications, the miniaturization of electronics, aeronautical advances, and knowledge about our own planet that couldn't be obtained from the ground. Questions about what NASA had been doing since Apollo, revealed a disconnect with the public over the purpose of government spending for human spaceflight. Public concerns centered on what we'd achieved since beating the Soviets to the Moon and at what cost.

I did my best to defend the program, espousing the usual rationales of international prestige and inspiration, but after the accident those justifications were wearing thin. I learned from my early experiences that defending the government's funding of human spaceflight in media interviews often required deftly avoiding landmines. I was determined to give honest and meaningful answers to what I was being asked, and dug more deeply into the issues.

NASA's justification of ancillary products referred to as "spin-offs," often seemed specious. If the government wanted to seed innovations like memory foam or cordless power tools, there were better ways to go about it than spending billions of tax dollars sending astronauts to space. The economic argument was also a bit deceptive, since direct funding of government jobs slows rather than stimulates the economy if it doesn't stimulate new markets.

In my view, the primary long-term rationale is simple. Humanity's only chance for survival as a species is to expand beyond the confines of Earth. This is a multigenerational goal and isn't NASA's sole responsibility. But the space pirates were already coming up with ways to utilize resources in space that could help people both on and off Earth.

I was fascinated by a book by Frank White called *The Overview Effect* that described how astronauts' view from space transformed their perspective about Earth's environment and humanity's ability to live and work together on our home planet. Every astronaut I'd met had shared how seeing the thin line of the atmosphere and land masses without borders changed their world outlook. This was certainly special, but I recognized it wouldn't make much of a difference until it was experienced by many more people, from all backgrounds.

For me, the value of human spaceflight is in its ability to transform humanity and society. One of my favorite examples of this power is the photograph called *Earthrise*, taken by the Apollo 8 astronauts from behind the Moon. The photo is one of the most famous of all time and is widely credited with starting the environmental movement. Robotic spacecraft had taken photos of Earth from space before, but it took a person to see the beauty of that unique perspective. Knowing it was the first time humans had seen the view with their own eyes gave the photograph more meaning to the rest of us.

I developed a reputation as a thoughtful and clear communicator about the space program and began to realize there might be a consequential role for me in the field. Assuring future space activities fulfilled their full potential to make a positive impact on society became my mission. I'd been planning to go to grad school to get an MBA or law degree, but I decided instead to pursue an advanced degree more directly aligned with my newfound goal and passion.

George Washington University offered a Master's program in International Science and Technology Policy, with a focus on space policy, and I attended night school while continuing to work full-time. The curriculum focused on history, and I enjoyed learning how lessons from the past could be adjusted to advance more effective policies and capitalize on the vantage of space. It sometimes bothered me that what drew me to the space program was different from what attracted nearly everyone else in the aerospace world. Instead of allowing myself to feel like a square peg in a round hole, I tried to think of ways I could connect the gears. I wanted to fill a missing piece of the puzzle to help the brilliant engineers and scientists solve the mysteries of the universe and advance civilization.

The Challenger accident was a game-changer that shook everyone involved in the space program to their core. NASA's ill-fated decisions exposed both poor management and technical failures that had been ignored by its renowned safety and engineering leadership. Less obvious at the time was how that cold day began the shift toward policies that would eventually allow the private sector to enter the arena in more significant ways. Whether it was fate or failure that kept me from taking calculus and pursuing engineering, as it turned out, studying policy and economics gave me a unique perspective that underpinned my career for the next thirty-five years.

2.

STAR STRUCK

I WAS PROMOTED TO BE THE EXECUTIVE DIRECTOR OF THE NATIONAL SPACE Society a year after the 1987 merger. Beyond making payroll, my biggest early objective was finding a way to marry the organization's previous cultures—the NSIers and L5ers. The differences between the two seemed to outweigh their similarities; their governance, history, and space ideology were polar opposites.

I discovered that while we had two different root systems, they fed off the same springs, the early successes of NASA and the vision of science fiction. Von Braun had recruited a board of renowned individuals from these fields in the 1970s, including astronauts Alan Shepard, Harrison Schmitt and John Glenn, and science fiction authors Ray Bradbury, Isaac Asimov, Gene Roddenberry, and Arthur C. Clarke. I did my best to tap them all as resources. Hugh Downs, a well-known television reporter, had already agreed to chair a Board of Governors, so I focused on recruiting a president and chairman who could also appeal to a wider audience.

As part of NASA's effort to have the Shuttle make space travel more routine and affordable before the Challenger accident, corporate representatives had been allowed to join astronaut crews if they accompanied commercial payloads. Charlie Walker had flown on the Shuttle three times between 1984 and 1985, as an employee of McDonnell Douglas. He was revered by the space pirates for pioneering citizen space travel, and I asked him to be our president. He agreed and became one of my earliest mentors. We worked together hand in glove for the duration of my time at NSS.

Having successfully recruited a stellar president, I set my gaze on finding a chairman, determined to shoot for the Moon.

I'd met Buzz Aldrin at a conference the year before and recognized that his perspective aligned well with the Society. As the second man to walk on the Moon, Buzz was an iconic household name and hero. I knew his involvement would garner public interest, increase our membership, and open doors. He was a rare early astronaut who preferred to focus on the future instead of the past. To my surprise, he agreed to serve as our board chair during our first phone conversation. He would hold the position for more than a decade.

Having Hugh, Buzz, and Charlie on the leadership team helped activate the other prestigious board members. It was an eclectic club of enormously successful people from a wide range of technical and nontechnical fields who shared a common bond of interest in creating a more valuable space program. As fascinating as it was to get to know them as individuals, the dynamic between them was even more compelling. I looked for ways to exploit the board's willingness to engage with each other that could help support the Society and the space program.

The merger was necessitated by a lack of financial stability in both organizations, so I involved the board in fundraising. I was seven years old when the original *Star Trek* series was canceled, but even as a latecomer I recognized the value of the brand Gene Roddenberry had created. I called him up to introduce myself. During my first visit to Paramount Studios, he got lost while driving me around the lot in a golf cart to meet both the original cast, who were filming their movie on one part of the lot, and the *Next Generation* cast who were filming their TV series in another. He let me sit in on his review of the day's taping, known as the dailies, and explained that he hadn't started out to write a show about space. Gene was a humanist and knew the messages he wanted to deliver would be more easily accepted in a futuristic setting. He was a fellow space pirate who was focused on what space exploration could offer to humanity.

We hit it off from the start, and Gene agreed to host a fundraising dinner for NSS on the *Star Trek* set for major donors. Nichelle Nichols and other *Star Trek* cast members even signed on to film public service announcements. Gene once told me that he created the character Wesley

Crusher to be what he envisioned as a perfect son—giving the boy genius his own middle name. Gene Wesley Roddenberry died in late 1991, while my husband and I were expecting our first child. When our son was born—after running it by Gene's widow, Majel—we named him Wesley.

Another coup was getting Gene to bring nearly the entire cast of *Star Trek: The Next Generation* to Washington for an NSS fundraiser in honor of the 30th Anniversary of Alan Shepard's flight. In addition to a large dinner, we held a star-gazing party at the National Observatory and a reception at Vice President Dan Quayle's residence, located on the same grounds. The evening was filled with stargazing—just not the kind that required a telescope.

Building on the event's success, we planned additional anniversary celebrations over the following years to raise money and expand public awareness of the value of space activities. Buzz agreed to headline dinners on both the East and West Coasts in honor of the 25th anniversary of the Apollo 11 landing. Vice President Gore and Dr. Carl Sagan spoke to space elites at our Washington, DC, dinner.

Although my second child was due less than two months later, I packed my not-so-flattering maternity formal and flew to Los Angeles the next day to roll out the red carpet for celebrity actors and astronauts at our second event. The Apollo 13 film was preparing to start shooting that summer, and I'd called up Universal Studios to invite the film's cast and crew to the gathering. I pitched it as a way to have them meet the real Apollo astronauts. It was a bit of a shell game because I also used the prospect of meeting Ron Howard, Tom Hanks, and the others to astronaut Jim Lovell and his crew. As I hoped would happen, both sets of invitations proved irresistible.

Jim Lovell expressed his disappointment during dinner that he had never received the Congressional Space Medal of Honor. The comment was made in passing, while he was explaining to the table that at the time of Apollo 13, the mission was considered a failure and NASA had done its best to sweep it under the rug. Hanks caught my eye and we both silently acknowledged the mention.

Several of the stars spoke that evening, but Lovell's comedic wit was especially memorable. Playing the perfect straight man, he explained that when he heard the Apollo 13 film was being made, he thought Kevin Costner should play him, since "they both had blond hair, blue eyes, and were still married to their first wives" (Costner and his wife divorced later that year). Lovell referenced several leading roles where Costner had portrayed strong men, such as *Bull Durham, Robin Hood, Field of Dreams,* and his recent Academy Award–winning performance in *Dances with Wolves.* The Costner bit went on for a while before Lovell, looking at Ron Howard in the audience, exclaimed, "Really, Ron? Tom Hanks doesn't look anything like me. You know there is finally going to be a Captain Lovell doll, but it's not going to look anything like me." Hanks was cracking up; this was his kind of humor. Lovell then went through a litany of roles Hanks had played that he felt were less worthy by comparison, including characters in *Bosom Buddies, Splash,* and *Big. Forrest Gump* had been released two weeks before, and when Lovell closed by saying his biggest concern was that Hanks might revert into the role of the intellectually challenged character while playing him, Hanks was hooked. It was their own meet cute and *they* became bosom buddies.

Hanks pulled me aside later that evening and said he'd noted my reaction to Lovell's comment about not receiving the Medal. He told me that if I was ever able to make that happen, he'd like to come to DC to see it be presented. The wheels in my head started turning as I considered how to maximize the value of this offer for the Space Society and the space program—and for Captain Lovell, of course.

NASA's Space Station program was in trouble on Capitol Hill. While it had the support of the usual suspects with aerospace jobs in their districts, that was barely enough to garner a majority of votes for its growing funding needs. After years of delays and escalating costs, a congressional effort to terminate the program had failed by only one vote in 1993. Attracting a wider base of support was critical to its future, and I wanted to tap into Tom Hanks's unmatched broad appeal.

After accomplishing the pressing task of delivering my second child, I put another shell game in motion, this one with even higher stakes. I suggested to Hanks's people that the medal ceremony at the White House was in the works and asked if he'd also speak at an event on the Hill while he was in DC. Tom Hanks agreed.

I pocketed his agreement, so the proposal to NASA and the White House to have President Clinton award Captain Lovell the medal came with the promise of Tom Hanks's attending and supporting NASA on the Hill. As I hoped, it was an opportunity none of them could pass up. After months of coordinating logistics, the ceremony was scheduled. NASA had agreed to host a small dinner in Captain Lovell's honor the night before and invited Senator Barbara Mikulski, who was already NASA's most important appropriations ally in Congress.

A few hours before the dinner, we received the message from White House staff that the next day's medal ceremony would have to be scheduled for a later date. I was told, "It doesn't fit the message of the day." I was still fuming when I arrived at the restaurant. Everyone else had taken it in stride, and Hanks had already changed his schedule to fly out the next morning–which meant he would not be giving his speech on the Hill. In my upset state, I shared the information with Senator Mikulski, who was seated next to me. We had kept the medal ceremony private, so this was the first the Senator had heard of it, and she agreed it was a missed opportunity.

During dinner, the maître d' brought a message to Senator Mikulski that she had an important phone call. This was before cell phones, so she stepped out of our private room to take the call. When she returned, she said it had been President Clinton, and she apologized for the intrusion. I noticed a twinkle in her eye when the maître d' interrupted a few minutes later with a phone call for Captain Lovell. He returned with a huge grin and news that the medal ceremony was back on the schedule for the following day. Tom Hanks gave me a thumbs-up, signaling he was changing his plans to stay as well.

Senator Mikulski told me later in the evening that the President had called to talk to her about the escalating situation in Bosnia, and after that discussion, she mentioned that her dinner companions, Jim Lovell and Tom Hanks, were sorry they wouldn't be seeing him the next day. The President claimed to have been unaware of the planned ceremony or the last-minute change and, as she relayed it to me, didn't seem very happy with the decision. Senator Mikulski didn't share her pivotal role with the others, but she is the one who made it happen.

When I arrived home from dinner, I had a message on my answering machine from my White House contact. It was late, but I returned the call and got an earful about how I'd overstepped many layers of White House protocol. The staffer ended the conversation by shouting that "this was not the way to get this done!" I didn't point out that the staff's cancellation meant a direct appeal to the President was our only option.

The medal ceremony was my first time in the Oval Office or meeting President Clinton, and I suffered from what has since been described to me as Oval Brain. Your first time being in the room is often so overwhelming that it becomes challenging to recall what happened afterward. I have pictures of myself shaking hands with the President and standing next to Colin Hanks (who had joined his father in Washington), watching Captain Lovell be presented with the medal, but I have very little memory of the experience. I learned to make mental notes in future Oval Office meetings to avoid having similar blank spells.

Tom Hanks's speech on the Hill later in the day, however, is still etched in my mind. It was an impassioned and heartfelt dissertation on the importance of the space program and the Space Station to a packed audience that included top aerospace brass and members of Congress. Hanks's remarks are still the most moving defense of the Space Station that I've ever heard. Although I had provided talking points to his staff in advance and was tempted to accept the credit others tried to give me, the speech was much more eloquent than my suggested notes.

Holding high-visibility activities impressed the NSS board and industry donors but was only a part of my strategy as executive director. I began to recognize that the reason there was lower public support for NASA likely went beyond just the message or the messenger. I believed it was related to the purpose of its human spaceflight programs. Movies about NASA's past achievements and future potential exploits captured the public's imagination because the storyline was meaningful. Blockbuster movies weren't made about astronauts circling Earth on a quest to learn how a few more of them could do it for longer periods of time. In national polls that asked citizens to list their highest-priority government programs, NASA sat near the bottom alongside foreign aid.

As I'd found after the Challenger accident, the public wasn't clear as to what the Agency was currently doing in human spaceflight, or why. Elevating NSS's reputation in the space community gave us the ability to convey a long-term purpose for space development. NASA's narrow focus on the handful of elected leaders obsessed with preserving NASA jobs in their districts was in contrast to the Society's vision of creating a spacefaring civilization that would establish communities beyond Earth. Congress had the power of the purse, but I learned in grad school that presidents had historically led policy shifts at NASA. I decided that the greatest opportunity to make a positive difference was to impact the space policy of future administrations.

Consistent with my personal political beliefs, I volunteered as a space policy advisor to presidential candidate Michael Dukakis in the 1988 election. Dukakis didn't end up giving Vice President George H.W. Bush much competition, but I wanted to do what I could to make sure a Democratic administration would have a worthwhile space program if he were elected. The experience gave me my first taste of policymaking at a national level, since my involvement with John Glenn's campaign in 1984 had been purely operational.

Space issues were managed as a subset of science policy on the campaign, so I attended meetings of the science policy group in DC. NASA wasn't central to the discussion, but when Senator Lloyd Bentsen from the space state of Texas became the vice-presidential candidate, it offered an opening to advance a positive space agenda. Governor Dukakis traveled to Houston with his new running mate in August of 1988 and announced his support for a "permanently manned space station" and promised to reestablish a Cabinet-level National Aeronautics and Space Council with the VP as Chairman. I was astonished by how easy it had been to get such an important commitment from a presidential candidate as a young volunteer, and I never forgot the lesson.

Pirates aren't known for working together well, and over time the NSS lost people on both ends of the ideological spectrum. Those who considered our vision unconventional joined or started more traditional industry associations and those who wanted a revolutionary approach created more radical organizations. But we were early thought leaders and a force for good in the space community. I was regularly invited to testify on Capitol Hill about NASA's budget and frequently contacted by the media to provide commentary on space issues. These opportunities allowed us to fill a void in space policy by articulating a valuable long-term purpose for human spaceflight. We were singing from the same sheet of music as Newt Gingrich, then-Minority Whip in Congress. Long before he became known for his malaprops as House Speaker, he was known for supporting space settlement. More importantly, when President George H. W. Bush was elected, it became clear his administration shared our views.

• • •

As the first President Bush took office, the space agency was still licking its wounds from the Challenger accident and had just returned the Shuttle to flight. Bush did what candidate Dukakis had promised and reinstated the 1960s practice of having a National Space Council chaired by the Vice President. Although he was not known as a heavyweight by the public, Vice President Quayle made the most of his assignment. Not having a space or technical background allowed the Vice President and his staff to be less influenced by traditional interests, and they seemed determined to set a new course at NASA.

In honor of the 20th Anniversary of Apollo on July 20, 1989, President Bush announced the Space Exploration Initiative (SEI), a program that would return humans to the Moon and go on to Mars. The President stated that "we must commit ourselves to a future where Americans and citizens of all nations will live and work in space. And today, yes, the US is the richest nation on Earth, with the most powerful economy in the world. And our goal is nothing less than to establish the United States as the preeminent spacefaring nation."

This was the first such presidential announcement I'd watched in person, and it felt historically important. The United States now had a stated national goal to become a spacefaring nation. The President of the United States said *spacefaring*! The Space Society quickly drafted letters for Buzz to send to the membership and initiated recruiting drives to build on the momentum of the President's speech. Key to the effort was activating our grassroots networks to have them contact their elected leaders, and we worked in coordination with the National Space Council staff.

The Space Council asked NASA for a ninety-day study to propose the new SEI program and made it clear they weren't looking for the same old way of doing business. They wanted NASA to look at a creative approach that might even require changes to the currently baselined Space Station Freedom program. As rumors leaked from the Agency about what NASA planned to put forward, the council asked a team from the Department of Energy's Los Alamos Laboratory to propose their own design for a program as a counter. The Space Council was clearly frustrated with NASA's lack of progress in human spaceflight and hoped that giving them some competition would inspire them to think differently.

President Bush had appointed former astronaut and Admiral Richard Truly as NASA Administrator not long before, and the Vice President and Council staff already seemed to regret his selection. In Quayle's autobiography *Standing Firm*, the chapter "Rockets and Red Tape" depicts his frustration over NASA's reaction to the White House's leadership as unrepentant. "The problem for NASA bureaucrats, a very pampered bunch," he writes, "was that space policy would be run by the White House.... They wanted to keep making space policy themselves, even if it was obvious to outsiders that the projects they had going were too unimaginative, too expensive, too big and too slow. NASA didn't mind having us help fight its budget battles with OMB and on the Hill, but they wanted to make up those budgets themselves."

The similarities between what the Vice President and National Space Council found in the early 1990s and what I experienced twenty years later are numerous. Vice President Quayle describes an early meeting with the NASA leadership when the head of space flight acknowledged they had been lying to Congress about the initial operating date for the Space Station program and even suggested they could fabricate technical reasons for what they already knew would be at least a four-year schedule slip.

Vice President Quayle recalled the rest of the delegation disassociating themselves from the remarks, but he said it was part of a pattern at the space agency that he and others in the executive branch found disturbing. "The arrogance was unbelievable," he writes. "They were just used to throwing around figures and estimates and counting on the old NASA glamour to dazzle whoever was listening. After the meeting Darman [the Director of OMB] told Albrecht [the Executive Secretary of the Space Council] that this made Watergate look small."

The Energy Lab submitted a proposal that included a series of inflatable habitats for a space station, lunar, and Mars bases, for a total cost of $40 billion. A few weeks later, NASA released its own proposal at a cost of $500 billion. The program was designed around NASA's existing infrastructure, and people. Instead of reshaping projects to the President's articulated goal of becoming spacefarers, it was a rehashed version of Apollo for even more money. The price tag drew headlines in the media, skepticism from Congress, and the ire of the Space Council.

The space pirates were thrilled with President Bush's announcement of a program to return to the Moon, most especially the Space Council's interest in creating a more innovative and sustainable program to do so. We shared our message in speeches and Hill testimony, and through grassroots lobbying efforts. Reading the transcripts of our testimony from over thirty years ago shows the space pirates' consistency of purpose:

> The best way to increase performance and lower cost over the long term is through a combination of government-supported exploration, government and private research and development, and routine private provision of space goods and services in a free-market competitive environment wherever possible.
>
> ...Even where government [was] the main purchaser, it should, as much as possible, act as a commercial customer—though perhaps with an eye toward rewarding innovative technologies that promise large cost reductions over time as opposed to old technologies that are price-competitive only because research and development costs have already been amortized.

We went on to explain that "in the space context, this would at the very least involve the use of result-oriented as opposed to hardware-specific contracts, and the abandonment of government micromanagement and detailed product specifications," reasoning that "such practices tend to inflate costs and to inhibit technological innovation—and, in the context of NASA, to tie up technical talent in contract oversight that should instead be employed in cutting-edge research." We urged the government to "become a more reliable customer where long-term procurements are involved" and concluded by saying that "NASA should do what it does best—advanced technology research and development—and let the commercial sector do what it does best—lowering costs on existing technologies as a result of the incentives provided by market forces."

If I were asked to give testimony on the topic today, I wouldn't change a word.

The clarity of our message was unique at the time, and Vice President Quayle and his team seemed to appreciate having a citizen organization in their corner. In addition to hosting our "star party" at his residence, he included us in meetings and events he held at the White House and at space conferences where he spoke. As a result of our increased visibility and clout, traditional aerospace interests started paying more attention.

The Society had an aerospace industry council made up of ten companies that each donated $50 thousand a year, which was not an insignificant percentage of our budget. Representatives of the council met quarterly and, in exchange for their tax-deductible contribution, were invited to our events and listed on our masthead. NSS space policy decisions were intentionally kept at an arm's length from the industry council in our governance model. NSS continued to favor increased NASA budgets and the Space Station, but our honed message was focused on supporting the new Space Exploration Initiative.

Boeing was a major player in NASA's existing programs and its representative Elliot Pulham, who chaired the council at the time, called me to levy a threat over our support for the President's new initiative. He said that if we didn't stop advocating for SEI and return our focus to Space Station Freedom, they would withdraw their financial support. I apologized for any misunderstanding about their role in our policymaking decisions and explained why we supported the new initiative. I then clarified that their tax-deductible corporate contributions were not allowed to be tied to specific programmatic advocacy, and I reminded him that this kind of "influencing" was unethical and against IRS regulations. My response was unexpected.

The industry council had backed me for the executive director position. I'd appreciated their confidence, but now I sensed their endorsement had come with strings attached. I was appalled by their blatantly expressed corporate views against those of the President, and their insinuation that they could control my actions. Not willing to be pulled off course, we continued activating our members and encouraged them to share their views with their elected leaders. Reports that letters were flooding Capitol Hill offices were confirmed when I received another disquieting call, this time from Kevin Kelly, a powerful Senate appropriations staff member.

The angry senior staffer said the number of calls and letters coming into their offices had become disruptive and unmanageable. He told me to "call off my dogs" or risk lower appropriations for the program. I recognized the bullying tone and responded by telling him what I knew to be true: the calls weren't going to stop, even if I were willing to do what he was asking. I explained that the individuals were simply exercising their rights as citizens, and I noted that they represented the view of people who cared about the long-term future of space development. Hill staff was accustomed to controlling corporate lobbyists with such tactics, and once again, my response was not expected. I held my ground, but the sand was shifting under my feet.

I had worked hard to establish positive relationships with people in the aerospace industry and on the Hill. I recognized that an attack on my reputation would not only hurt my career but also could diminish support for the Society. It was an early firsthand lesson in the tactics of traditional lobbyists, bureaucrats, and Congress. I tried to take it more as an obstacle than a verdict, believing that sound policies and principles would prevail.

The executive secretary of the Space Council, Mark Albrecht, later penned a book about his experiences working with NASA at the time and expressed his own frustration. He noted that it was "often said that the Pentagon is an iron triangle of industry, the Congress and the military services. In fact, the civil space program is a steel quadrangle of industry, Congress, the NASA bureaucracy, and academic scientists. In the end, there was little left of what was once the crown jewel of the age of American exceptionalism." The Space Council had also considered the opposition to SEI as more of an obstacle than a verdict, but they ultimately lost their battle.

It took a few years, but I developed positive working relationships with Elliot and Kevin. The Space Exploration Initiative didn't fare as well. The President had requested $200 million in its first year, which Congress almost entirely eliminated. SEI limped along for a few more years with tepid support and small amounts of funding, but it never moved beyond conceptual mission studies. It would not be the last time that a sitting president wasn't able to overcome the status quo and the formidable combination of self-interested industry lobbyists, the Hill, and NASA's entrenched bureaucracy to achieve a more innovative and sustainable space program. I would be at the center of the storm twenty years later and should have been better prepared for the deluge.

After putting up with his foot dragging and half-measures, the Bush administration decided to send Admiral Truly packing from his role as NASA Administrator. Vice President Quayle was given the assignment to tell him, but when Quayle asked for his resignation, Dick Truly said he worked for the President and would need to hear it from him. On February 12, 1992, soon after firing the astronaut admiral Administrator that he had appointed less than three years before, the President told the Vice President and his Space Council chief that he wanted a brilliant new Administrator confirmed by April 1. It was a nearly impossible task.

The space community was troubled by the dismissal of the astronaut Administrator, even when word spread that he had not left gracefully. Industry colleagues were disappointed, and even the trade press seemed to side with the Admiral against the President. I was disheartened by how many of my long-held beliefs were being shattered. The combination of aerospace support for government handouts over sustainable space development, a political appointee disregarding the President's direction, and a president waiting more than two years to dismiss someone for such transgressions, undermined the values and tenets I held for both space and politics.

• • •

Against all odds, the National Space Council delivered on President Bush's directive. Daniel Saul Goldin was sworn in as NASA Administrator on April 1, 1992. He was brilliant and new. Dan was relatively unknown to the Washington civil space establishment and didn't make it a priority to join their club, knowing he was hired to drive a transformative agenda. He'd spent the first five years of his career working on electric propulsion for NASA in Cleveland before being recruited away by the aerospace firm Thompson Ramo Wooldridge, Inc., the company better known as TRW that had developed the lunar lander engines for Apollo. What Dan saw in industry was light-years ahead of the space agency, and he realized even then that NASA's arteries were beginning to harden. Years later, he told me that the people who worked on Apollo were terrific, but the government system did not allow for innovation, and NASA's bureaucracy was stifling.

Dan arrived at NASA at an inauspicious time, just eight months before a presidential election that put the continuity of his politically appointed position at risk. Even so, he got right to work addressing what he viewed as a mandate for positive change. With the backing of the National Space Council, he addressed many early challenges. On the top of his list was tackling the problems of having too many support contractors and too much bureaucracy.

The 1992 election made George H. W. Bush a one-term president, so in January he started dutifully packing his office.

The NASA transition team for President-elect Clinton was headed by Dr. Sally Ride, and Dan directed his staff to work with her to gather information that would help prepare a new team. It wasn't much of a secret that President Clinton wanted Ride to be his NASA Administrator, but it was also pretty obvious that she didn't want the job. In mid-January, much to Dan's surprise, the President-elect's staff asked if he was willing to stay until they could find a new Administrator.

Sally had been willing to support the three-month-long transition but continued to demur at the White House's attempts to recruit her for the Administrator position. Hoping that she would change her mind, they were willing to wait her out.

Dan Goldin made the most of the extra time. Delays and increased costs had diminished congressional support for Space Station Freedom, so he looked for a way to bring the Democratic Congress and new President aboard. The Russians had more experience with human spaceflight and space stations than anyone, but with their economy in ruins following the breakup of the Soviet Union in 1991, they were struggling to keep their own space station, called Mir ("peace"), operating.

President Bush had begun cooperative astronaut exchanges with Russian cosmonauts flying on the Shuttle and US astronauts flying on the Soyuz—the Russian space capsule—to Mir. Dan proposed a soft power diplomatic tool, namely expanding this post-Perestroika effort to include an invitation to the Russians as full partners on a redesigned space station. The pitch resulted in a home run at the White House, giving both Dan and the Space Station the backing they needed from the new Clinton administration.

Acknowledging the significance of the revision, the Space Station Freedom name was changed to the International Space Station and became known as ISS. Dan also eliminated the layer of support contractors and bureaucrats who had been managing the program in favor of a more streamlined management structure.

The fourteen existing Space Station partners and industry contractors were understandably upset about the program changes and complained that their views hadn't been given due consideration. The decision meant reorienting the station to a higher inclination to accommodate launches from Baikonur, which would in turn require upgrades to the Shuttle. The changes also meant scrapping nearly eight years of progress and 400 tons of new hardware already in development. Less acknowledged in the debate was that without adding the Russians, Freedom wasn't likely to survive the chopping block.

Even with President Clinton's support, Station funding passed in the House of Representatives by only one vote in 1993. Dan's bold move to restructure the program and invite the Russians was taken at considerable risk, but in hindsight, gave the Space Station a more enduring geopolitical purpose. When the Columbia accident grounded the Shuttle a decade later, Russian access to the ISS quite literally saved the program.

In addition to human spaceflight, several other important science and technology programs were fundamentally restructured under Dan's leadership. As he arrived, NASA was working through a fistful of embarrassing failures, most notably, the loss of Mars Observer, and the blurred vision of the Hubble Space Telescope. These losses cost taxpayers over $2 billion and were more visible and disruptive because they increased the incentive to add marginal requirements, while reinforcing the vicious cycle of larger, fewer missions. When costs rose, innovation declined, since it was hard to justify adding new technologies when it meant risking billion-dollar, once-in-a-lifetime missions.

Transitioning to missions that were "faster, better and cheaper" had been proposed by the Bush Space Council, and became Dan's mantra. The concept was to have lower cost and more frequent missions that could test more innovative technologies. Dan instituted competitions for science missions at a range of costs that still exist today. Even these changes drew fire from the space industrial establishment, since traditional programs funneled more money to universities and contractors in key congressional districts and by design were harder to cancel.

My professional path crossed with Dan Goldin often during his first few years as Administrator, including when he helped secure the Space Medal of Honor for Jim Lovell. He was a space pirate at his core and believed in the message I was delivering on behalf of the NSS. In 1994, Dan appointed me to serve on the NASA Advisory Council, the prestigious board previously made up of older, distinguished white men. Two years later, he called to offer me a job helping him with strategy at NASA, and I didn't hesitate to say yes. I'd been at the National Space Society for over twelve years and was anxious to put what I'd learned into practice.

My first year at NASA was a challenge. Dan was known as a manager who thought people worked better out of their comfort zones. Every day seemed like a new test. His leadership team didn't much care for someone new being dropped into their midst to do strategy, especially someone without a military or engineering background. Dan's top consiglieres, Mike "Mini" Mott and Jack "Zorro" Dailey, both former marines, were especially hostile to my presence. Mini and Zorro had a more traditional agenda than the Administrator and knew I'd been hired to offer a dissenting opinion. They both did their best to keep me from succeeding. Information is power, especially in government work where salaries are nearly equal, and it seemed they thought that by sharing very little with me, I'd be unsuccessful and fly the coop.

One of the few other senior women at NASA, Deidre Lee, was the head of procurement and had previously led acquisition policy at the Department of Defense. She befriended me early on and became a mentor. Dee referred to these male colleagues as the cup boys. The reference was to their ubiquitous coffee mugs adorned with their military call-signs: Mini, Zorro, Dragon, Panther, and so on. It helped to have someone else acknowledge the exclusionary culture, and the moniker stuck. I've worked with many cup boys throughout my career and found their predisposition to oppose new ideas and new people was often contrary to NASA's mission.

Mini and Zorro's intent to clip my wings eventually gave me flight, but not as they intended. Cutting me out of daily correspondence and meetings gave me time to focus my attention on the strategic issues that mattered most, and led to my promotion to run NASA's policy office.

Observing Dan taught me to envision a goal and work backward, often referred to as right-to-left thinking. This lesson wasn't something he taught; it was simply how he operated. Lots of people talk about being strategic, but few people make it an everyday practice. Dan is wired to have the endgame at the forefront. People drawn to work in a bureaucracy tend to be process-type thinkers, so having a strategic leader is vital. I saw plenty of people focused on the process of designing, developing, and operating missions, which was not at all my skill set. I also noticed that end-state goals were whittled down in a project's development cycle, even as costs went up. The system needed fundamental change to break this pernicious cycle, but very few government leaders were focused on transformative solutions.

"Faster, Better, Cheaper" is most often associated with the revolution Dan executed at NASA. The truth is, enabling commercial practices also owes much to his leadership. As a policy person, and with what I'd learned from the space pirates, I decided this area was the most constructive place to bend my pick. I formed a group that recommended ways to drive commercial opportunities in human spaceflight.

With Dan's blessing, I recruited a team that included people from both the program and general counsel's offices and enlisted an astronaut with firsthand experience in conducting experiments on orbit, Dr. Mary Ellen Weber. Together, we plowed the fields and planted seeds, advancing commercial policies that helped provide a blueprint for successes that followed.

The team first took on facilitating early commercial utilization of the International Space Station. NASA had invested tens of billions of dollars in the development of the facility, yet budgeted only a few hundred million dollars for its use. Dan wanted to change this paradigm. He spearheaded an agreement with the National Institutes of Health to develop groundbreaking biological science as well as stimulate private-sector investment and used his power of bully pulpit to attract potential new users. *The New York Times* highlighted NASA's intent to allow privately funded experiments on the Space Station, and Fisk Johnson, a great-great-grandson of the founder of S. C. Johnson Wax, read the article. He contacted NASA expressing his interest in the concept, and Dan assigned my team to follow up.

Fisk was interested in developing and funding a meaningful commercial scientific experiment on the Space Station. He'd grown up excited by the space program and was an adventurer, pilot, entrepreneur, and environmentalist who wanted to make a difference. Our effort led to a cooperative agreement to test metabolites on liver and kidney tissue in space. Fisk paid NASA several million dollars to fly the experiment to the ISS in 2001.

Working on the project was an incredibly valuable opportunity, and NASA was lucky to have such a competent, motivated, high net-worth individual as a partner. The NASA team was top-notch, but even so, everything took longer than expected. The experience helped educate us about usage of commercial practices and introduced me to the unique partnership authority that became critical to lowering the costs of space transportation ten years later.

Unfortunately, the project also demonstrated the challenges of undertaking commercial experiments in a laboratory where the researchers weren't present. On orbit, the astronaut assigned to conduct the experiment made a critical timing mistake when injecting the metabolites in the liver cells, negating any potential results. NASA offered a no-cost re-flight, and the astronauts on STS-107 eventually conducted the experiment two years later. The results were lost along with the Columbia and her crew as they returned to Earth on February 1, 2003.

The second Shuttle disaster was as disruptive to the future of human spaceflight as the first. The Space Shuttle had escaped cancellation after Challenger with nearly three years of painful, excruciatingly expensive re-engineering intended to restore faith that the underlying concept was sound. Columbia tore that facade apart. Now there was no denying that the Shuttle was never going to live up to its intended goal of making space transportation cheap or reliable.

In reaction to the first Shuttle accident, the government had paid the two remaining launch companies, Lockheed Martin and Boeing, to evolve and operate the rockets they'd developed for the military in the 1960s. It was an attempt to create a competitive system, but it became more profitable for the companies to focus on the guaranteed government market, where they could keep prices high. Over time it only disincentivized them from competing in the commercial market. Launching on the newly Evolved, Expendable Launch Vehicles, known as EELVs, wasn't as expensive as launching on the Shuttle but was well above the rest of the world's prices and drove the commercial satellite business to French, Chinese, and Russian rockets.

Not being competitive in the commercial launch market left the government footing the entire bill for both Shuttle and EELV systems, costing taxpayers billions of dollars in direct payments. Sustained higher launch costs were also a disincentive to developing more innovative, cheaper satellites. Trying out newer, lower-cost technologies on a satellite added risk that was harder to accept when the launch alone cost a few hundred million dollars. The expense of both systems reduced risk tolerance, limited incentives to innovate, and stifled potential competition, reinforcing a vicious cycle.

A few years later, when US government launches alone weren't enough to sustain business for even two companies, the threat that either or both rocket systems could be discontinued led the government to support combining them into a single company, creating a sanctioned monopoly. The joint venture was named the United Launch Alliance (ULA), and it secured an annual billion-dollar subsidy for institutional overhead, on top of ever-increasing launch prices.

ULA describes its formation as a response to the government's desire to consolidate, but such characterizations are usually more insidious. Members of Congress who represent aerospace workers are often told by the industry what it needs to maintain or increase jobs in their districts, and when the government complies, the industry says it is acting on the government's request. Once large contracts for jobs and infrastructure are awarded in key congressional districts, it becomes nearly impossible to change course. Dan and others have referred to this system as the giant self-licking ice cream cone.

An increased financial burden to the taxpayer is just one of the adverse outcomes from this self-feeding cycle. The system eliminates the proven method for driving efficiency and innovation—competition. Without competition, not only do costs increase, but the incentives are reversed. Members of Congress and industry who have perfected the system of lapping up all the ice cream for themselves understandably enjoy the sugar high, but over the longer term, it undermines the health of the sector and the nation. I like ice cream as well as anyone, and without the disincentive of it making me fat, I too would overindulge more often.

Forty years into the space program, the most significant barrier to realizing the full potential of space was still the cost of getting to orbit—escaping gravity. Dan referred to this barrier as the Gordian Knot, the key to unlocking sustainable space development. Even within a flat budget and faced with increasing Shuttle and ISS costs, Dan committed resources to begin loosening the knot.

NASA's typical procurement mechanism, similar to the military, establishes the details of what it wants to purchase (a set of formal requirements) and solicits bids for either fixed-price or cost-plus contracts. Federal Acquisition Regulations (FARs) encumber contracts with detailed restrictions and requirements but give the government control, insight into the contractor's work, and any intellectual property that is developed as a result.

Nearly all of NASA's large contracts are managed as cost-plus, since companies are not well suited to shoulder the risk of unknown increases associated with one-of-a-kind programs. These are referred to as *traditional* contracts at NASA, but there is nothing traditional about the model in the consumer world. The contracts allow for a negotiated cost, plus a guaranteed fee and whatever extras might be needed to cover any new requirements or unexpected challenges that crop up during the project's development. Since NASA's appropriations are annual, overruns are accommodated by slipping schedules, so contractors most often get paid even more money for longer periods of time.

Fixed-price FAR contracts are typically limited to smaller or strictly defined purchases that can be precisely scoped with fewer unknown variables. Companies under fixed-price contracts still manage to get paid more than what's in their contract; if things don't go as planned or the customer wants changes, the government's only recourse is to walk away, or renegotiate for a higher price and longer term.

Neither procurement mechanism is appropriate for programs focused on developing services or capabilities designed to serve additional customers, which is the goal of developing lower-cost space transportation. Dan asked the lawyers to find a better way to achieve this unique goal and they proposed moving forward with a cooperative agreement instead of using the traditional procurement process. Partnerships can be used for third parties to conduct research and develop specific systems, but the government is not allowed to dictate the solution. Cooperative agreements are managed outside the FARs, and are typically lower cost to the government. This is the same structure we used to partner with Fisk Johnson on his Space Station research experiment.

Dan initiated a large competition for government-industry partnerships called the Reusable Launch Vehicle (RLV) program in 1996. The largest RLV program was a test demonstration vehicle called X-33. Its stated purpose was "to build a vehicle that takes days, not months, to turn around; dozens, not thousands, of people to operate; with launch costs that are a tenth of what they are now." NASA's investment was to incentivize development of a commercial reusable launch vehicle that would cut the cost of getting a pound of payload to orbit from $10,000 to $1,000. Three major aerospace companies completed Phase One of the program and submitted their own designs to develop a test-vehicle in Phase Two. This phase was to fund a demonstration mission and required cost-sharing proposals that would lead to fully commercialized systems.

The competition was NASA's first significant commercial partnership for a reusable launch vehicle—the holy grail of lowering launch costs. The program was strongly supported by the Clinton administration, and having interest from the top aerospace companies on the project earned it initial bipartisan support in Congress. Most of the space pirates supported McDonnell Douglas's proposal, known as the Delta Clipper or DC-X, but Phase Two was awarded to Lockheed Martin in 1996 with a goal to have a first flight of a sub-scale, suborbital vehicle by 1999. Lockheed planned to have an operating vehicle, which it called Venture Star, by 2005. I led a study in NASA's policy office that evaluated incentives that could assist private sector partners in developing the fully commercialized follow-on vehicles

The X-33 award was made two weeks before I began working for NASA, a point I am quick to clarify, since my husband ended up working on the program for Lockheed Martin. Dave's employment with the aerospace industry giant has lasted over twenty years, a reality of our relationship that has surprisingly caused very little stress between us. Having met him on John Glenn's campaign, Dave's space interest preceded my own. We got married on Space Day—July 20, the anniversary of the Moon landing—and named our first child after the creator of *Star Trek*, after all.

Our boys have been steeped in space events and conversations their entire lives. They were two and five when I started working at NASA. I struggled with the typical difficulties of being home less often than most of their friends' moms, coping and compensating in all ways possible. Our youngest was particularly upset when I was leaving for a work trip one weekend, and in the pre-9/11 days, he was able to join me on the NASA plane to "see what Mommy did" before we took off. The pilots showed Mitch the cockpit; then he went into the cabin to check out the big soft seats and play with the phone in the armrest. The snacks in the galley soon caught his eye, and when he opened the mini refrigerator and saw it stocked with sodas, he looked at me with his big brown sad eyes and said what I recall being his first sentence, "It doesn't look like work." From the mouths of babes.

David and I have been careful to adhere to conflict-of-interest rules throughout our careers, which became more complex in my later NASA employ. On the RLV program, my team's policy work related to potential government incentives for any partnership program, so it was deemed not to be a conflict by NASA's General Counsel.

Our analysis showed how incentives such as loan guarantees and service purchase agreements, or "anchor tenancy," could help private partners finance their development costs, required because the government wouldn't be providing direct funding for the full-scale vehicle. The concept of anchor tenancy was that the government would buy goods and services, instead of building or operating a system themselves. It was based on the successful Kelly Air Mail Act of 1925, which stimulated early commercial aviation by giving airmail contracts to airlines. Once fledgling airlines had a base of government funding secured, they could search for private sector customers at more reasonable price points and start building an industry.

The cost-sharing partnership for the X-33 test program was fixed-price to the government, so all cost overruns were borne by the company. Lockheed Martin invested over $350 million of their own money and NASA spent $900 million over the four years of the program. When the test vehicle experienced technical challenges, instead of adding the estimated $50 to $100 million to the program, it was terminated. The X-33/VentureStar program never came close to launching, but even its veiled threat to more traditional interests helped send it to an early grave. The partnership approach was dismissed along with the program, but the development of the commercial cargo and crew program several years later owe much to this early effort.

In addition to the X-33, the RLV program included a smaller variant called X-34, which evolved into a military space plane to fly classified space missions. Alternative Access to Space (Alt Access) partnerships were funded to support start-up companies, many led by space pirates. These were inspired by the same growth projections for the satellite market that drove X-33 bidders' interest in developing their own launch vehicles. Four start-up companies received initial funding through Alt Access before being reduced to one by the next NASA Administrator.

The most enduring transportation policy of Dan's drove cost savings in the procurement of launch services instead of vehicles for NASA science missions. Purchasing launch services reduced NASA's costs and freed up government resources to do more scientific research. It also enabled nascent companies to develop sustainable business cases that could attract customers beyond the government. The policy thread woven into the Gordian Knot began to loosen.

While Dan invested NASA resources to drive down launch costs, the space pirates continued to make progress in other ways. In the spring of 1996, one of my space pirate friends, Peter Diamandis, needed support for a project he was creating called the X-Prize. Designed to spur the development of fully reusable spacecraft that could carry passengers to and from the edge of space, the project was based on the concept of the Orteig Prize that Charles Lindbergh had won in 1927. The X-Prize was to be awarded to the first team to build a space transportation system that could carry a human (plus the ability to carry two more) to space and do it again within two weeks. Peter called me to ask if I could help him get the project endorsed by NASA.

There is no better way to demonstrate the advantage of right-to-left thinking than offering a prize for results. This concept has been demonstrated throughout history, and I became an immediate proponent. I knew the project aligned with Dan's ideology, but the problem was convincing his handlers. The cup boys felt the risk of a NASA endorsement was too high, given that people could potentially die attempting to win the prize, and they encouraged Dan not to support the project.

The bureaucracy overplayed their hand. This was the exact type of hardening of the arteries Dan wanted to correct. He accepted the X-Prize invitation to attend the kick-off event in St. Louis, honoring the city where Charles Lindbergh began his own flight, which signaled NASA's endorsement. Dan's early support for the X-Prize was pivotal to the advancement of what became a major contributor to eventual private spaceflight, even though it took longer than many of us expected.

The X-Prize envisioned a winner in five years, so by 2000, time was running out and Peter reached out to me again for assistance. Dan had become discouraged by the project's lack of progress and didn't want to stick his neck out again. I was happy to stick my own (much less important) neck out, and did what I could to use my position as the head of the NASA policy office to help. I hosted a meeting with Peter and his bankers to signal NASA's continued endorsement, which helped secure temporary financing for the Prize. The funding kept the doors open until Anousheh Ansari came to the rescue. Her donation formally established the $10 million dollar Ansari X-Prize in 2004.

The space pirates and projects like the X-Prize continued to make strides toward reducing the cost of human spaceflight. NASA's leadership didn't prioritize the use of private partnerships or reusability to lower launch costs for human spaceflight after Dan left, even after the Columbia accident. Eight years later, when I returned to the Agency in 2008, I picked up the ball not much farther than where Dan had left it on the field.

Dan worked tirelessly to battle the status quo, advance innovation, and transition NASA into the twenty-first century for nearly ten years and under three different presidents from both political parties. I owe a debt of gratitude to many people who took a chance on me throughout my career, and Dan is high on that list. The space pirates helped shape my space ideology, but it was Dan who forged it into steel.

3.

MODERN MYTHS

FIFTEEN YEARS AFTER THE FIRST MOON LANDING—WHEN I WAS JUST STARTing my career—the frustration in the space community over a perceived loss of public support was palpable. Having won the race to the Moon, NASA had earned a reputation as one of the most preeminent and revered organizations of all time. This exalted status made it even more difficult to accept that before the celebration was over—as the last astronaut walked on the Moon in 1972—NASA's budget was only a bit more than half what it had been at its peak. The Agency wanted to take on similarly audacious missions, but no new national purpose could justify the cost.

Our stalled progress in human spaceflight is typically blamed on a lack of political will, but this thinking overlooks what garnered that political will—beating the Russians. The audacious goal was designed to address the greatest threat US leaders perceived at the time, and NASA delivered brilliantly. Without an opponent to race after the Moon landing, the political will to fight communism was understandably invested elsewhere, leaving NASA all dressed up with nowhere to go. Apollo's unique mandate drove strategic and technical decisions with no regard to lowering operational costs that would have led to a more sustainable program.

Remembrances of Apollo focus on the bold, daring dreams of a young president and romanticize the era and purpose of the mission. "We choose to go to the Moon in this decade and do the other things, not because they are easy, but because they are hard" is a justification still repeated in an attempt to increase NASA's funding today. The historic record is clear, but still this narrative prevails, reinforced by historians and institutions invested in perpetuating the legend. We all want to recreate a time when our nation seemed pure and good. Mythology sells.

One-dimensional time periods and mythological characters are created purposefully to deliver a message, but true stories and people are multidimensional—a combination of holy and unholy motivations. Recently released tapes of Kennedy's views on space expose a more complex narrative. Outside the rhetorical flourishes of his public speeches, the Kennedy recordings reveal that within a year of proposing the mission, he questioned its value. Rarely acknowledged are the recordings of Kennedy telling NASA Administrator James Webb in November 1962 that if we can't beat the Russians, "we ought to be clear, otherwise we shouldn't be spending this amount of money, because I'm not that interested in space." It is widely recognized that political leaders convey different motives for decisions in public speeches than they do in private conversations. Even so, listening to the recording of President Kennedy tell the NASA Administrator he doesn't care about space is jarring, given the myth of Camelot.

Concerned about rising costs, JFK made several serious offers to Soviet leader Nikita Khrushchev to cooperate on lunar astronaut flights in the hope of reducing the expense. In hindsight, we can see that the Apollo myth obfuscated even President Kennedy's *public* statements, such as his 1963 UN speech where he proposed a cooperative lunar program with the Russians.

"Why," he asked the audience, "therefore, should man's first flight to the Moon be a matter of national competition?" Kennedy noted, "The clouds have lifted a little," in terms of US–Soviet antagonism. "The Soviet Union and the United States, together with their allies, can achieve further agreements—agreements which spring from our mutual interest in avoiding mutual destruction."

By then, the Bay of Pigs invasion and the Cuban Missile Crisis had come and gone.

Back in April 1961, when President Kennedy had asked Vice President Johnson, who led the National Space Council, to "give me a goal that I can win," it was in reality the pinnacle of success for a self-interested community that had its sights set on sending people to the Moon decades prior. Their timing could not have been better.

On April 4, 1961, President Kennedy gave the go-ahead for the covert military invasion of Cuba that was executed ten days later. The failed Bay of Pigs invasion embarrassed the Kennedy administration and increased his need to show strength and leadership against the Soviets. When the Russians successfully launched Yuri Gagarin into space on April 12—in the middle of the debacle—the trap was sprung. Having lost to the Soviets in the race to launch the first satellite and man to space, NASA had to offer something far enough in the future that would give the United States time to catch up. The NASA boss took the advice of his top rocket scientist, former Nazi officer Dr. Wernher von Braun, and recommended a manned lunar landing.

It was exactly what the young President needed at that moment—a bold anti-communist vision. Kennedy gave his historic address to Congress that announced the Apollo program less than a month later—May 25, 1961. At that point in history, the week I was born, the United States had precisely 20 minutes of experience in human spaceflight under its belt. Never mind that we had lost the old race; a new race had begun.

Kennedy's decision was in some ways preordained by what had transpired since the Russians launched Sputnik three and a half years earlier. In her groundbreaking podcast titled *Moonrise*—released for the 50th Anniversary of Apollo—Lillian Cunningham from *The Washington Post* reveals recently released transcripts and recordings that describe how NASA's link to the Cold War was purposefully intensified by self-interested parties. Individuals under the leadership of von Braun proactively worked to link space exploration and national security to exploit the opportunity to increase funding for their own projects.

Research conducted by Margaret Mead in the weeks following Sputnik portrays a very different public reaction than what became part of the American zeitgeist. In personal surveys taken immediately after the launch, Mead found many Americans only moderately interested and not overtly hysterical about the beach ball–sized satellite. Any initial tepid public reactions to Sputnik were quickly fueled into a blaze of paranoia by the military-industrial complex, politicians, and the media—the typical beneficiaries of a frightened public.

President Eisenhower was at Camp David when he was informed about the Russian satellite, and he didn't even opt to return to Washington. The launch had been expected, after all. Documents released by the CIA in 2017, sixty years to the day after the launch, state that "US intelligence, the military and the administration of President Dwight D. Eisenhower not only were fully informed of Soviet planning to launch an Earth satellite, but also knew a Soviet satellite would probably achieve orbit no later than the end of 1957." The President even sent a congratulatory message to the Soviets and was privately relieved at not being first. The President and those around him welcomed the launch to help establish the principle of "freedom of space," the idea that outer space belonged to everyone, thereby allowing satellite flights over foreign countries.

Eisenhower's lack of a strong antagonistic response to Sputnik played into the hands of those who wanted to coerce the nation into a space race. Fiercely dedicated to averting nuclear war, President Eisenhower was concerned that funding space stunts would limit funding for Inter-Continental Ballistic Missiles (ICBMs) that in his view were much more critical to national security. Parochial and partisan interests depicted him as passive and unconcerned, with a goal to inspire a more heightened reaction from the Democrats. These special interests—including von Braun—prodded then-Senator Lyndon Johnson to further exploit the opportunity. As Chairman of the Senate Preparedness Investigation Subcommittee of the Senate Armed Services, Johnson was urged to hold an "Inquiry into Satellite and Missile Programs." Months-long hearings began in late November 1957, and they included witness testimony from seventy-three people in support of increased space activities.

Longing to create the future they envisioned, scientists, bureaucrats, and science fiction writers—several of whom would become my future colleagues—provided quixotic, unrealistic testimony about the expectations of space exploration. They found their mark in LBJ, exploiting his penchant for embellishment; in his closing statement at the hearing he said, "Control of space means control of the world. From space the masters of infinity will have the power to control the Earth's weather, to cause drought and flood, to change the tides and raise the levels of the sea, to divert the gulf stream and change temperate climates to frigid." Historians agree that the hearings were critical to garnering support for the creation of a space agency.

When the first US attempt to launch a satellite suffered from public failure, newspapers ran headlines such as FLOPNIK, KAPUTNIK, and STAYPUTNIK, and the Democrats doubled down on their criticism of the administration. As Eisenhower found himself on the losing end of the space race, he had no choice but to rely on one of the prime instigators behind popularizing and politicizing space—Wernher von Braun—to launch America's first satellite and try to change the narrative.

After the successful launch of Explorer One on the last day of January in 1958, von Braun became a national hero, and he and his supporters continued to lobby for a new cabinet agency for space. Eisenhower came up with an alternative, less powerful independent agency, evolving the organization from the National Advisory Committee for Aeronautics. After months of negotiations with Congress, President Eisenhower signed the NASA Space Act in July of 1958, which took effect on October 1.

The politics that undermined Eisenhower's reputation on space issues were closely related to his greatest concern about the future of the country: the potential abuse of power of the expanding government system related to the armaments industry. Eisenhower was not the only leader concerned. Ralph Cordiner, an industrialist and businessman who was CEO of General Electric from 1958 to 1963, wrote in 1961:

> We must recognize that there are growth tendencies in these government agencies that could over expand under the pressures of the space program, unless proper safeguards are established. As we step up our activities on the space frontier, many companies, universities, and individual citizens will become increasingly dependent on the political whims and necessities of the Federal government. And if that drift continues without check, the United States may find itself becoming the very kind of society it is struggling against—a regimented society whose people and institutions are dominated by a central government.

As he left office in 1961, Eisenhower chose to focus his final speech from the White House on his concern about the growing power of the military-industrial complex:

> In the councils of government, we must guard against the acquisition of unwarranted influence, whether sought or unsought, by the military–industrial complex. The potential for the disastrous rise of misplaced power exists and will persist. We must never let the weight of this combination endanger our liberties or democratic processes. We should take nothing for granted. Only an alert and knowledgeable citizenry can compel the proper meshing of the huge industrial and military machinery of defense with our peaceful methods and goals, so that security and liberty may prosper together.

After serving forty-six years in the military and government, Eisenhower recognized that this force had already solidified into a permanent armaments industry. His efforts to keep a civilian space agency separate from this growing threat were in recognition of the problem, but only partially successful. The perceived threat of Soviet domination after WWII, propelled by the self-interested industry, led the United States to the Korean and Vietnam Wars and many other failed interventions. It also fueled the civil space program.

America's soft-power effort to prove the superiority of our democratic system became part of the playbook to beat back communism. Stimulated by this rationale, NASA's annual budget increased from $2 billion in 1960 to a record $34 billion in 1966, and it shot us to the Moon. Choosing to create a big socialist program to win the Moon Race succeeded but had negative consequences that have also been papered over by historians caught up in the legend.

While space historians universally agree that the justification for the original founding of NASA and human spaceflight was the Cold War, they rarely question whether this linkage was valid. Beating the Russians to the Moon in 1969 was an amazing achievement, but it didn't end the Cold War. *That* would take twenty more years. According to the renowned Cold War historian Archie Brown, there is no direct link between beating the Russians to the Moon and the eventual fall of the Soviet Union. It is possible that US achievements in space gave pause to a handful of nations that were considering closer affiliation with Russia, but no new countries renounced their ties to the USSR after the Moon landings. Again, distance, context and perspective give us a more complete view.

In their book *Merchants of Doubt: How a Handful of Scientists Obscured the Truth on Issues from Tobacco Smoke to Global Warming*, historians Naomi Oreskes and Eric M. Conway document how a handful of "right-wing ideologues" have (mis)shaped US policy for decades, delaying government action on life-and-death issues from cigarettes and secondhand smoke to acid rain, and now, to climate change. Out of four scientists prominently featured in the book, I worked with three of them—Robert Jastrow, Frederick Seitz, and Fred Singer—after they'd been recruited by von Braun to the NSI (predecessor to NSS) board. They had been key players driving the Cold War narrative to Lyndon Johnson and subsequently JFK and Eisenhower. Their effort helped convince both Soviet Premiers and US Presidents that dominating space was the most meaningful measure of a Cold War superpower. Self-interested parties make similar claims today about being in a race to the Moon against China—ignoring the reality that we've already won.

A full measure of history can only be understood by the passage of time. Even then, it is shaped by those who do the telling.

Wernher von Braun's fundamental role in the development of rocketry, NASA, and human spaceflight is well documented and universally acclaimed. I spent the first twelve years of my career working for the organization he founded in an attempt to increase public support for the space program. I sat at his former desk in an office with a large, stunning photograph of him on my wall. Sitting under a portrait of von Braun gave me mixed feelings. Although he was the brilliant father of the space program, he was also a rightly vilified Nazi SS officer.

Von Braun's narrative is perhaps the most carefully crafted myth of all. Not only was he a leader in the Nazi Party, von Braun created the V-2 rocket that caused over 20,000 deaths—9,000 from attacks, and 12,000 caused by the forced participation of laborers and concentration camp prisoners. The V-2 wasn't developed to carry bombs; von Braun wanted it to carry men to outer space—to the Moon. Interviewed about it later, von Braun said, "The rocket worked perfectly, other than landing on the wrong planet." Being forced to make a choice between receiving the resources to advance his rocket or being killed was his understandable justification, but even that defense doesn't give consideration to the lives of enslaved Jews or victims of his weapons.

Public awareness of von Braun's role in WWII was parodied by popular musician Tom Lehrer in a 1965 song that included the lyrics:

> Once the rockets are up, who cares where they come down,
> that's not my department says Wernher von Braun.

By that time, the rocket designer was a revered member of one of the most exclusive clubs in the country, NASA. Few would disagree that von Braun was a brilliant man who changed the trajectory of the American space industry, but the full truth of his story has too often been whitewashed. The ends may have justified the means to von Braun when his life was on the line, but choosing to ignore his role in killing innocent civilians, not to mention idolizing him as a person, point to the sense of superiority and singlemindedness by some within the NASA family.

As the armaments industry saw their business decline and they turned their focus to a new, more lasting enemy—the Communists—von Braun was an important and willing ally in achieving their interests. Within fifteen years of his surrender to the United States, von Braun was leading much of NASA and advising our political leaders. It is hard to imagine that today, twenty years after the attacks on the World Trade Center, our government would allow anyone at all affiliated with the hijacking to play a central role in US technology programs or consort with national leaders—even if their participation in the attacks had been coerced. There is little doubt that von Braun's persona as a dashing white man with blue eyes and blond hair contributed to his speedy assimilation into the United States, NASA, and the halls of power in Washington.

Nostalgia for the early "manned" space program depicts it as a period that was thrilling and wonderful for everyone. In reality, that was primarily the case for Anglo-Saxon white men. I don't know many women or minorities who harken for the days when we didn't have the right to vote, join country clubs, or get credit cards without our husband's permission. I love the fashion and hairstyles in the *First Wives Club* and *Mad Men* as much as anyone, but I have no desire to return to eras when only male careers were dominant and a single woman making it out of the secretary pool was viewed as sufficient progress.

In the 1960s, NASA—a civilian space agency—was tasked with essentially a military objective as an instrument of the Cold War. This linkage increased NASA's budget immensely but also drove the fledgling space agency's culture toward building and operating its own large engineering projects and away from more universal investments in technical innovation and scientific research. The massive institutional bureaucracy and industry interests developed for Apollo required exorbitant fixed costs just to be maintained. Once in place, legacy interests were naturally conditioned to seek missions and goals that could use the same infrastructure and similarly motivated workforce. The space-industrial complex became a victim of its own success.

4.

RISKY BUSINESS

WHEN THE COLD WAR FINALLY ENDED AT THE CLOSE OF 1991, NASA DEMONstrated its ability to be nimble and opportunistic—two qualities not typically associated with bureaucracies. I've often cited this example when asked why I thought NASA could ever evolve to embrace commercial companies launching their astronauts to space: they let their former mortal enemy do it.

The fall of the Soviet Union hit their space program hard, and sensing an opportunity, US space policy leaders quickly pivoted to supporting peaceful cooperation between the programs. Our goal was to sustain what had been high-tech jobs in areas outside the military—modernized swords into plowshares. After Perestroika, discussions of joint missions with astronauts and cosmonauts on the Shuttle and on Mir began in late 1992 by President Bush and were continued by President Clinton. Eleven missions were conducted from 1995 through 1998 under Dan Goldin, who used the program as the basis to propose inviting the Russians to become full partners on NASA's planned space station.

Pulling back the Iron Curtain to view our competitors' program was of great interest to the United States and NASA, so these initiatives were not entirely benevolent. The Space Station Freedom program had already received more than $10 billion in its first ten years with no launch in sight. NASA hoped to gain insight, knowledge, and much-needed hardware from the Russians. The former Soviet Union had a tremendous capability but needed an influx of Western currency. Thus, the deal was struck.

In a truly historic irony, the Russian Space Agency (RSA) turned to capitalism to fund its space program and began selling tourist seats on

Soyuz for trips to its Mir Space Station. Not only was NASA partnering with its former antagonist, who had been responsible for its very existence, but the Russians were adopting the ideology their own space program had been created to discredit. Meanwhile, NASA remained stuck in a system based on centralized planning.

Russian commercial space activities had begun a few years earlier, encouraged and facilitated by several early space pirates, including Walt Anderson and Jeff Manber, who formed a company called MirCorp in 1999 to privatize the Russian Space Station. The company offered wealthy individuals and corporations visits to the Mir. The first space tourist, Dennis Tito, reportedly paid the Russians $20 million to travel on a Soyuz to the ISS in April 2001, through another early space tourism company called Space Adventures.

I left NASA at the end of the Clinton administration and was working at an aerospace consulting firm in the summer of 2001 when an opportunity came along that gave me a front-row seat to the early days of the Russian space tourism business.

Fisk Johnson, the S. C. Johnson's Wax heir I'd worked with at NASA, reached out to engage me in facilitating his own visit to the Space Station. As a pilot, scientist, and entrepreneur in his early forties with the means to purchase a seat, Fisk Johnson was an ideal candidate and client. His interest wasn't in getting a joyride or publicity; it was to train and conduct the scientific experiment he and his team had developed over the previous five years.

I had traveled with Dan Goldin to Russia for a Soyuz launch a year before and met some of the key players at the Russian Space Agency. I also knew the leader of MirCorp and was able to negotiate a significantly lower-than-advertised price for my client to become the third space tourist on ISS. I accompanied Fisk and his small team to Russia when he began his medical certification that summer.

The Institute for Biological and Physical Problems—known as the IBMP—conducted cosmonaut medical certification for RSA at a nondescript facility in Moscow. A few of the tests took place in Star City, the cosmonaut training center, toward the end of the process—if you made it that far. Fisk performed well on his medical certification and completed all the tests with high marks in only a few weeks. The team finalized the details for the Soyuz flight with the full support of MirCorp. The ten-day mission we negotiated for was to launch in the fall of 2002. Our agreement called for six months of training, which would be spread over the next year in order to accommodate Fisk's other commitments.

We were all back in the United States before September 11, 2001, when the hijacked planes hit the Twin Towers in New York City and the Pentagon in Washington, DC. I was in my top-floor office across the street from the White House when we first heard of the attacks. Several of us headed to the roof to take in the scene for ourselves. We saw streams of people running out of the White House complex and then noticed the billowing smoke darkening the sky in the direction of the Pentagon. Realizing it was not a drill, we ran down the stairs and joined the throngs of people already running up Connecticut Avenue, away from the White House, which we feared was the target of another attack. I was in heels so didn't make it very far before hearing that the fourth plane had crashed in Pennsylvania. I borrowed a pair of tennis shoes from a friend who lived nearby and headed to my suburban home on foot. The view of a smoldering Pentagon as I ran over a Key Bridge empty of traffic is forever etched in my memory.

The events of 9/11 changed many things, including Fisk's ability to spend six months of the next year training for his Soyuz mission. Like others, he needed to focus on his business, which had been disrupted and needed his attention. When I called Jeff Manber at MirCorp to break the bad news, he asked if I knew anyone else who might be able to pay for the seat. The Russians' ability to fulfill their commitment to the ISS depended on receiving Western dollars from tourist flights.

Being in Russia with Fisk had laid bare that nation's economic challenges, and it was clear that safely maintaining the production of the Soyuz was at risk. Without a regular infusion of cash, the future of human spaceflight seemed to hang in the balance. I felt somewhat responsible, since it was my client who had backed out. I asked a few high-net-worth people who'd previously expressed interest in flying on the Space Shuttle if they might be interested in purchasing the seat. James Cameron was too tall for the Soyuz; Tom Hanks was waiting for his kids to get older, and Leo DiCaprio was ... too busy. I began to consider more creative options when I heard the back-up plan for the seat was to fly an astronaut from the European Space Agency, which paid even less than the contract we'd negotiated for Fisk.

My NASA policy work had included overseeing a branding study a few years before that had uncovered a significant interest in private-sector marketing related to human spaceflight. It suggested that consumer brands like Nike and Disney were willing to pay to be associated with the space program, but as a government agency, neither NASA nor its employees—the astronauts—could endorse commercial products. I contacted the firm that had done the study and asked if they thought raising sponsorships to pay for a tourist to travel on the Soyuz to the ISS was viable. They not only said yes but suggested it would ideally be a mom. A woman would get earned media from being the first female space tourist, and mothers made 70 percent of household purchasing decisions and were therefore favored by sponsors.

Not willing to pass up the unique opportunity, I built a proposal around flying myself and signed with an agent. My objectives for the project included increasing public awareness of the value of human spaceflight, conducting Fisk Johnson's experiment that had the potential to utilize ISS to design life-saving medicines, validating commercial space practices, and getting funding to the Russians so they could fulfill their commitments to NASA. The goal of my space career was never to fly in space personally, but to fundamentally open space. Getting to go personally would be icing on the cake. It wasn't without risk, of course. But I knew the Soyuz was the safest way to travel to space, and my family gave me their support. My consulting firm also signed on and helped design the project. We named it Astromom and reached out to MirCorp to begin a new round of negotiations.

The next eight months were an intense and surreal combination of negotiating, planning, recruiting sponsors, being interviewed by the media, attempting to learn a bit of Russian, and completing my medical certification in Moscow. Initial discussions with the Russians signaled we could get the seat for $12 million, and our early sponsor recruiting indicated raising that amount of money was doable.

We selected the Discovery Channel as the project's media partner, and they agreed to pay $500 thousand for exclusive rights for three television episodes focused on training, the flight itself, and my return. With this negotiated but unexecuted agreement in hand, the team built a portfolio of interested sponsors in the low-million-dollar range that included Disney, Sudafed, Major League Baseball, and RadioShack.

Most sponsors wanted to film commercials of me using their product or services in space. Disney wanted a video of me at the landing responding to the question, "Lori Garver, you just went to space, where do you want to go next?" Sudafed had been used by astronauts in space to clear their nasal passages for years, and this was finally their chance to brag about this important endorsement. Major League Baseball wanted me to throw out the first pitch from space, something they did with astronauts years later for free. We never finalized an agreement with MLB, but it got serious enough that I signed both boys up for baseball instead of soccer that season. (One of them has never forgiven me.)

Negotiations were underway for the named, primary sponsor in the $3 to 5 million range: it would either be Visa or Mastercard. The Soyuz was scheduled to land in November, so the concept was for me to buy my kids Christmas presents on the mission, which would be the first credit card purchase in space. RadioShack was interested in being the store to sell me the gifts—another coveted sponsor opportunity.

The Winter Olympics in Salt Lake City was a target-rich environment for meeting with potential sponsors, and my agent invited the whole family to join, showing off our photogenic ten- and eight-year-old boys. I made appearances on *The Today Show*, *Good Morning America*, and national nightly news programs. On *The Daily Show*, Jon Stewart even featured a clip that included my boys—poking fun of course.

Other than the Discovery Channel and an initial small sponsorship from RadioShack, the agreements were proving difficult to finalize. The primary sticking point was the risk to the companies in case of a catastrophic accident. No one wanted their logo on a flight suit that might end up charred on the Kazakh Steppe. Not that the companies were ever so blunt. We were finding workarounds, such as delaying publicity around sponsorships until my safe return, so we decided I could begin my medical certification. All our discussions hinged on my qualifying for the flight; the timing was critical.

I knew what was in store at the IBMP and Star City, since I had watched Fisk go through the certification just a few months earlier. I'd always been a bit of a daredevil, enjoying speeding down hills on wheels or skis, but I hadn't fully appreciated the certification experience secondhand. Though I wasn't a natural, I eventually powered through it all in good health and spirits. Unlike my former client, who had stayed in a nice hotel, I stayed onsite in the no-frills, dorm-like facilities for the cosmonauts and bunked with my interpreter and her mother in a small city apartment on the weekends. This was space training on a budget. A classic mom quality.

In March 2002, I was in Moscow undergoing my medical testing when the television show TMZ announced that Lance Bass—a member of the boy band NSYNC—would travel to space on the Russian flight that fall. Neither MirCorp nor the Russian Space Agency had heard anything about Lance Bass or the prospect of him flying on the Soyuz. I stuck to my training and decided that if what the entertainment press was saying was true, I'd enjoy the company.

I mastered much of the testing, including atmospheric, pressure, cardio endurance, high-altitude, mental, and physiological. For each test, I was hooked up to wires through anode stickers placed on pulse points around my body. Doctors would monitor how I managed the different aspects of stress associated with the varied experiments. The most challenging of the tests for me was the vestibular training, which was essentially a spinning chair. I had watched Fisk pass the test on the first try, so I gave it a whirl. When my heart rate elevated and I began to perspire, the doctors could tell I would soon vomit. They pulled me out of the chair. I hadn't made it too far into my first test and would be given only one more chance to pass.

I continued with other medical procedures, including a full body x-ray bone scan, gastroscopy, and colonoscopy, all without any sedatives or anesthesia. Not only do you need to pass medically, but you also need to show you can manage yourself through extreme discomfort. For me it wasn't just physical but emotional discomfort. If it was easier for the doctors to have me entirely naked for a test, that is how it was conducted. No gown or sheets to cover me during x-ray, ultrasound, or gynecological exams. Halfway through a lengthy gynecological exam, my male doctor asked my interpreter if I was feeling pain. When I replied no, he smiled and asked, "Does it feel good?"—in English.

I was nervous about passing the vestibular test, so I researched strategies to get through it—including talking with former NASA colleagues who worked with astronauts on biofeedback. One of my Russian doctors told me she had a particular interest in seeing me succeed and suggested I think about what I enjoyed doing most: When was I happiest? I settled on the ritual of tucking my boys into bed. I asked the doctors if I could sing during the test, and they didn't see why not. I passed my final vestibular test singing a medley of songs by John Denver and Rodgers and Hammerstein, at a resting pulse.

Another stumbling block surfaced in a routine ultrasound when they found a gallstone that needed to be removed before completing my physical certification. The final test was to be the centrifuge, and nobody's allowed to undergo the required 8-Gs with gallstones. Given my experience with Russia's medical practices, I opted to go back to the United States to have it removed, planning to return to complete my tests and begin training in a few weeks.

By now Lance Bass had heard what the media was reporting as a *competition* between us, so he sent me a dozen roses and four front-row tickets to an upcoming NSYNC concert in DC as a peace offering. I reciprocated by arranging an after-hours private tour of the National Air and Space Museum, where I met Lance the next evening. We both ended up in Moscow a few weeks later, Lance to begin his medical testing, and me to complete my centrifuge test.

I did my best to support Lance's attempt, since his going to space would still have accomplished many of my own objectives—providing the much-needed funding to the Russian space program and increasing public awareness. But I knew if he could pay the advertised price of $20 million, there was no way I could compete. I went back to Russia to help facilitate his introduction to MirCorp, and since I'd already paid for the medical certification, I wanted to see it through to completion.

Lance and his team stayed in a trendy Moscow hotel while I was back on a cot in the IBMP dorm eating boiled eggs, sardines, and beets. I was proud of the contrast. His team included me in their social gatherings, and I enjoyed teasing Lance as he went through the myriad of tests I'd already completed. A few cosmonauts invited us to join them at a dacha in the country to practice our shooting skills one weekend. The trip was right out of central casting, complete with day drinking before heading out with our rifles to shoot skeet. No one died—which was a real concern at one point—so it became one more amazing life experience. Lance reported that it seemed like just a normal day growing up in Mississippi.

NASA had an astronaut representative stationed in Star City, and Lance asked if I could make an introduction. I knew Bob Cabana and agreed to facilitate a lunch meeting with the three of us. Bob asked Lance early in the conversation what he studied in school. When Lance explained he had to drop out of school to join the band, Bob asked what he'd been studying before he had to drop out. Lance clarified that he hadn't dropped out of college; he had dropped out of high school. Bob did his best not to appear shocked, but the remainder of lunch was awkward, and I could tell my own qualifications were starting to look better by comparison.

Our final test, the centrifuge, was conducted in Star City, and Lance and I were scheduled to take the test on the same day. In turn, we were each suited up with a slew of analogue sensors at our pulse points and headgear so every bodily function and brain wave could be analyzed. The goal was to keep your heart rate and perspiration as low as possible as they ramped up the g-forces. We also had to follow an instrument panel, pressing switches in reaction to a series of lights while they measured our response time. The test ended if your perspiration or pulse rate got too high, or if your reaction time slowed.

Lance went first, and I watched from the upper gallery as the long arm holding the orb that encapsulated him began to spin. I was standing with the operators and doctors, so I could also see his face through the mounted camera in the capsule. The team of doctors read the instruments and made notes on clipboards as one of them called out the increasing G-levels through a microphone. At about 7-Gs they pointed to his face on the screen and laughed. His lips and cheeks were flapping in response to the forces. The goal was 8-Gs, and Lance made the mark before they started throttling down the speed.

As he stepped out of the capsule with a big smile on his face, I nervously climbed aboard. My favorite doctor had given me more tips to pass this test and I took her advice. I was self-conscious, knowing Lance and the others would be laughing at my flapping face, but as the count reached 7-Gs, my focus was entirely on the lights and switches. My peripheral vision began to darken—the first sign there isn't enough blood in your head and you will soon black out. I managed to use my muscle tightening exercises and biofeedback to keep up with the flashing lights on the instrument panel to 8-Gs, and Lance was waiting as I climbed out of the ball. He gave me a big hug. Neither of us could stop smiling and laughing over the unique shared experience.

At the completion of each test, I typically met with the doctor to discuss my test results. Liquor and sweets accompanied the meetings, no matter what time of day. As Lance and I sipped cognac and ate cookies after the centrifuge test, the doctor looked over our results. Our reaction times were similar, but my heart rate had remained lower, which was most likely the result of my being a forty-year-old woman instead of a twenty-three-year-old, testosterone-filled man. I was naturally ecstatic and asked the doctor if Yuri Gagarin's doing well on the test had helped him be selected for his spaceflight. She looked at me and shook her head, responding in English, "No, he had the best smile." I looked at Lance and said what we of course already knew: that his beautiful smile was more likely to get him the flight over me, as well.

The final indignation from the Russian doctors was to have me stand naked in front of them as they looked at my full-body x-rays and discussed my anatomy. In my case, they were completely flummoxed that I didn't have back pain, based on a twist they saw in my lower spine. They seemed to talk about it endlessly, as my translator did her best to tell me what they were saying, before deciding it wasn't a problem for spaceflight. They recommended I never play tennis or go skiing again—advice I have ignored. Lance completed his last requirement on the same day, and we all celebrated our formal certification Russian style, with many toasts of vodka.

As I expected, the Russians saw dollar signs when Lance arrived in Moscow. The price for the Soyuz seat shot back up to $20 million. Lance hadn't been informed that he'd need to pay for the trip initially, so he was never planning to spend his own money. An agent unaffiliated with Lance had learned about my sponsorship model, and after reading in a fan chat room that Lance had a boyhood interest in being an astronaut, the agent decided to adapt it for him. Convinced my model would work but would benefit from the free publicity around someone already famous, the agent sent Lance a fax that "invited" him to go to space, even though no one had even talked to the Russians or started to look for actual sponsors. MTV eventually signed on as his media partner, but on a non-exchange-of-funds basis. RadioShack pivoted from their sponsorship of me to sponsoring the initial training for both of us, which we announced at a press conference together in Moscow in May 2002.

Lance's team had initially offered to pay for me to train as his backup, but within days MirCorp and RSA started complaining that Lance hadn't paid for his own training, much less mine. The media reported that he and his entourage had skipped out on the bill for their hotel stay before moving into more moderately priced accommodations. Training was going to be a few hundred-thousand dollars, and time was running short, so I accepted reality and returned home. It was a once-in-a-lifetime adventure, and the insights I gained into the Russian space program, firsthand experience raising commercial sponsorships for spaceflight, and the personal satisfaction I got from pushing myself physically and mentally were rewards that remain with me today.

I will never know if I would have been aboard the Soyuz that launched from Baikonur, Kazakhstan, on October 30, 2002, if Lance hadn't shown up and derailed my agreement with the Russian Space Agency. Had I flown, the experience would likely have been life-changing and advanced at least a few of my objectives for the mission. But even the small amount of training I completed made it obvious to me that I would not have made a great astronaut. Not surprisingly, the qualifications required to be a policy analyst don't necessarily translate to having the physiological stamina and aptitude for spaceflight. The reverse is obviously also true, but rarely considered.

• • •

After my stint as Astromom, I returned to my consultant business at the Avascent Group and enjoyed working in a world where all parties' agendas and incentives were aligned. The Astromom experience led me to another consulting engagement for two more potential space tourists, but that opportunity ended when the Space Shuttle Columbia disintegrated over Texas in early 2003, leaving the US government purchasing Soyuz seats for its own astronauts—at over $50 million a seat—to gain access to the Space Station.

The Columbia accident occurred just over a year into the tenure of NASA Administrator Sean O'Keefe, whom President George W. Bush had appointed to replace Dan Goldin. I did my best to steer clear of the new Administrator after hearing from one of my clients that an aide of his suggested that unless they stopped using me as a consultant, NASA wouldn't work with them. I hadn't even met O'Keefe at the time, so his attempt to blackball me was either partisan or based on my having worked for Dan Goldin. It was an unethical practice no matter his motivation, but I laid low and kept my list of clients under wraps during his tenure.

O'Keefe's three years leading NASA haven't received much scrutiny, and that isn't my goal here. We've only met a few times, and he has always been collegial. O'Keefe had been Secretary of the Navy for the last six months of the first Bush administration and previously served as Comptroller of the Department of Defense under Dick Cheney. O'Keefe had degrees in history and public administration and no background or experience in the civil or commercial space world. Nevertheless, he was welcomed by the space establishment.

The Space Shuttle Columbia lifted off on January 16, 2003, after being delayed eighteen times over a period of two years. It was the 113th Shuttle mission. Cameras captured footage of a particularly large piece of foam insulation breaking off from the fuel tank and directly hitting the leading edge of the orbiter's wing 82 seconds into the flight. The NASA team tracking the issue raised concerns when they reviewed footage the day after the launch. One mechanical engineer assigned to the Shuttle program described the risk in an email as having the potential to lead to an LOCV—NASA shorthand for the loss of the crew and vehicle.

A NASA engineering briefing assessing the potential damage to the orbiter concluded they needed images from spy satellites and made their request to senior managers. When no photographs were received, one member of the team wrote in a follow-up email to "beg" for imagery. A few days later, a Boeing analysis concluded Columbia could safely return even if there was significant tile damage, and NASA leaders accepted the conclusion without seeking any images.

Speculation about the reason the Agency's leadership didn't follow up on the engineers' request range from concerns that Department of Defense (DoD) resources were needed in preparation for the pending US invasion of Iraq that began a month and a half later, to a belief that if the damage was significant enough to be seen from a spy satellite, there wasn't much NASA could do about it anyway. In any case, who gave the direction not to seek assistance was, in engineering speak, never pounded flat.

On February 1, Columbia was over Texas and setting up for its usual landing approach into Florida when abnormal readings showed up at Mission Control. The NASA communications lead, known as Capcom, called Columbia on a private channel to discuss the issue. Columbia's Commander Rick Husband responded, "Roger," but whatever he tried to say afterward was indecipherable before the line was cut off. Mission Control received a call a few minutes later saying that Dallas television stations were broadcasting video of the Shuttle breaking up in the sky. The Flight Director ordered the doors locked and computer data saved. Search and rescue teams later in the day confirmed the astronauts had not survived the incident.

Although the Challenger investigation committee many years prior had been set up independently, Sean O'Keefe was allowed to internally commission the Columbia Accident Investigation Board (CAIB).

Admiral Harold Gehman, a retired four-star admiral from the Navy was appointed as its Chair. As it had with the Challenger accident, the investigation board found that senior management had ignored technical safety issues. NASA was aware that foam insulation from the external tank had come loose and fallen off during launches, often hitting the orbiter, but instead of fixing the problem, NASA had normalized the deviation. They decided it wasn't a problem because it hadn't been a problem.

Beyond technical factors leading to the tragedy, there were organizational causes. The CAIB concluded that NASA did not request photos because of bureaucratic confusion and management errors. The Board identified a pervasive attitude at NASA that the shuttle was "operational" rather than "experimental," and they discovered this attitude caused managers to enter decision-making with a "prove it's not safe to launch" rather than "prove it is safe to launch" mentality.

My first tour at NASA had fallen between the two Shuttle accidents, and I returned to the Agency six years after Columbia. As a NASA senior leader for more than ten years—eight of them while Shuttles were regularly flying—I've given a lot of thought to management's actions surrounding both accidents. The lessons I took away contributed to my belief that managing within the government system is often misaligned with the incentives for technical safety and success. In both Shuttle accidents, NASA's leaders were balancing a number of factors unrelated to safety that led to their fateful, critical decisions.

The pressure NASA felt to show Congress and the President that the Shuttle was economical and reliable contributed to the Agency's decision to waive the restrictions against launching the Challenger in freezing temperatures. The same pressures led NASA to ignore the ongoing incidents of foam loss that caused the Columbia disaster. Political conflicts related to requesting assistance from other government agencies may also have contributed toward the decision not to seek outside resources that could have given the astronauts at least a chance of survival.

Even fundamental decisions made during the Shuttle's development were more aligned with political interests than safety. Pressure to lower immediate costs and use existing infrastructure and workforce to garner

Congressional support in key districts led NASA to make design trade-offs, such as the use of solid rocket motors, which had previously not been considered safe for human spaceflight. Strapping the orbiter to the side of the rockets, instead of on top, was a decision that put the astronauts literally in the line of fire.

In the private sector, answering to shareholders and investors incentivizes against risky decisions that would "bet the company." Concerns that industry will cut corners without regard to safety are, in my view, mostly misplaced. As an example, commercial airlines in the United States fly 900,000,000 people a year and in the past ten years as of the writing of this book, those airlines have had two in-flight fatalities among their nine billion passengers. Flying in the environment of the lower atmosphere is a much less dynamic endeavor than flying through it to and from space, but the government's safety record of non-combat aviation-related fatalities is extremely poor compared to the airlines' solid record.

Over a dozen military personnel fatalities are the result of aviation accidents each year—a huge percentage difference, given the number of people flown compared to commercial aviation. In 2018, there were thirty-nine non-combat aviation-related deaths out of one and a half million active-duty military and reserve personnel. Accidental deaths have exceeded combat deaths in the last ten years. A similar accident rate for US airlines would equate to many thousands of deaths per year. History has shown that it is difficult for the government to inspect itself objectively.

A lack of independence from the government-appointed accident review board was raised as an area of potential conflict of interest related to the Columbia disaster investigation. When Congress expressed this concern, the NASA Inspector General (IG) submitted an unsolicited letter to the Hill saying he had concluded the board was acting independently and without "undue influence" from NASA. Unfortunately, this wasn't the first time this NASA IG had gone out of his way to defend the Agency head he was charged to patrol.

The Inspector General Act of 1978 defines the purpose of government agency Inspector Generals as providing independent audit and investigation functions to combat crime, fraud, waste, abuse, and mismanagement. NASA's IG had been replaced within three months of O'Keefe's arrival at the Agency by Robert "Moose" Cobb, who was widely viewed as having been chosen by the new Administrator—an atypical practice perhaps allowed because of his senior White House relationships. Cobb's close contact with O'Keefe, and other improprieties, fueled numerous investigations during his tenure.

A 2006 investigation found that "Cobb lunched, drank, played golf and traveled with [then Administrator] Sean O'Keefe and emails showed he frequently consulted with top NASA officials on investigations raising concerns about his independence." Three members of Congress—two Democrats and a Republican—urged President Obama to oust Cobb in 2009, saying the Inspector General "has been repeatedly accused of stifling investigations, retaliating against whistleblowers and prioritizing social relationships with top NASA officials over proper federal oversight." Cobb was forced to resign in 2009 and was replaced by a more well-regarded IG within months of my return to NASA.

The first twenty-five years of my aerospace career included jobs in the nonprofit, government, and private sectors. My closest professional associates were people at NASA, in the aerospace industry, on the Hill, and in Democratic and Republican administrations. I worked with space pirates, hero astronauts, Hollywood stars, pop stars, and the Russians. These experiences shaped my views about not just space policy, but about governance and management practices across the sectors.

Gaining a deeper understanding and respect for our accomplishments in space at times revealed behavior I viewed as self-dealing and unseemly, both in and outside of government. For me, NASA was teetering from its pedestal, giving in to gravity. Frustrated by the government's slow progress, many space pirates were developing their own advanced technologies and private initiatives. I believed they were on the right path, and I was determined to use my different skill set, knowledge, and experience to help NASA embrace a more positive and collaborative role in the future.

PART TWO

FORCE

def. Strength or energy exerted or brought to bear; coercion or compulsion; to make someone do something against their will

5.

LOOKING UNDER THE HOOD

DEMOCRACIES REQUIRE THEIR CITIZENS TO PARTICIPATE IN THE ELECTORAL process. It is a defining characteristic of representative government that shouldn't be taken for granted. When the 2008 presidential campaign got underway, I was determined to be personally engaged. I cohosted a fundraiser for Bill Richardson, who was an early believer in the value of commercial space, and when his campaign failed to gain traction, I branched out and attended events for Barack Obama and Hillary Clinton. They were brief but determinative encounters. Candidate Obama's reaction to my question about what future he saw for NASA was that he "wanted to do fewer things better." Although this seemed like a fair criticism, the same question to Hillary elicited a more fulsome answer and discussion.

I started volunteering for Hillary's campaign in May of 2007. As her lead on space issues, I developed policy documents, provided input on speeches, served as her space spokesperson, and represented her in surrogate debates on the topic. I ran two caucuses for Hillary in Iowa, and those icy-cold weeks knocking on doors there harkened back to my formative years campaigning in Michigan. I was wrecked when she came in third in Iowa and eventually lost the nomination to Barack Obama. Walking out of her concession speech in the DC Pension building—a venue selected for its intact glass ceiling—I wasn't ready to fall for Obama immediately. His "oh, you're likable enough, Hillary" line in one of the early debates still stung.

When the Obama campaign reached out to former Hillary volunteers, and we had our first substantive conversation about space, he quickly won me over. Obama and I share Midwest sensibilities and were raised with an idealistic view of public service. We were born in the same year, so grew up at a time when NASA's star was bright. Our similar ideology about the role of government had left us both unsatisfied with NASA's post-Apollo progress, and we shared a goal to revive the Agency.

Being less of a fangirl of Senator Obama, I wasn't as nervous around him as I had been around Senator Clinton—which proved to be extremely valuable. I felt free to speak my mind when he asked me if I agreed with "Ben" Nelson about extending the Shuttle. I didn't struggle for words when he asked what I would do instead. When I received the call a few weeks later asking me to lead the NASA transition team, I was eager to help. I thought that his unmatched ability to communicate and potential to transcend barriers just might be compelling enough to get him elected.

The pending retirement of the Space Shuttle put a review of the Constellation program front and center for our transition team. Begun by NASA Administrator Mike Griffin in 2006, Constellation was designed as a fully government owned and operated human spaceflight program, tasked to be both a replacement for the Space Shuttle and to send astronauts back to the Moon. There were several elements planned for the program, but only Ares I, a crew launch system; Orion, a crew capsule; and ground systems were funded in NASA's five-year budget runout. A much bigger rocket called Ares V, a lunar lander referred to as Altair, space suits, lunar vehicles, and other key requirements for lunar missions were aspirational, since there was no funding available until after the Space Station was de-orbited, which was scheduled for 2015.

When we discovered in 2009 that problems with Constellation had already pushed the launch of the first two elements (Ares I and Orion) to 2016—after the planned de-orbit of the Station—it wasn't really all that surprising. Like the Shuttle and Station, Constellation had been designed to utilize the infrastructure and workforce that had been built for the Apollo program. Being sized to use fifty-year-old existing, expensive facilities at their capacity in an attempt to gain political support was never going to be efficient. Maintaining decades-old facilities assured that not only infrastructure but personnel costs would remain high. Mike Griffin and others considered this a positive feature of the program, since it would satisfy key congressional delegations who would keep the money flowing.

Given the size and importance of Constellation to NASA's future, it should have been expected that many of the transition team's questions were focused on the program. Yet, NASA and contractor managers hunkered down and hid information about the program from us when we arrived at Headquarters in November 2008. Our briefings primarily centered on intricate artist renderings and high-def videos but were short on substantive details or answers to most of our questions. The attitude to keep us in the dark was pervasive throughout the program's management. Casual interactions with former colleagues in the hallways were viewed as suspicious by NASA leadership. The message conveyed from the top was that being seen even talking with us would be "career limiting."

One of the people who tried to get the transition team information about the program was Sally Ride. Sally was on the Board of the Aerospace Corporation, which had recently reviewed the Ares I rocket at NASA's request. She had been briefed on the initial results and thought we should see them. The Aerospace Corporation is a Federally Funded Research and Development Corporation that was founded in 1954 to advise the Air Force and other aerospace related agencies. Their reputation for excellence and independence was renowned, so we set up a briefing at their offices.

The first few slides of the presentation included boilerplate information about the organization; the Aerospace Corps team was going through them at a glacial pace. When they paused on about the fifth slide without relevant information, it appeared they were concluding the briefing. I asked if this was all they had for us, and they somewhat reluctantly acknowledged that was the entire presentation. We were incredulous. They had obviously been directed not to share the substance of the review with us, and the message had likely come from the top. The four of us walked out after fifteen minutes. I later confirmed with a colleague that a NASA directive had indeed been delivered.

At one of our few face-to-face discussions during the entire three months of transition, Mike Griffin expressed feeling insulted that our team was "looking under the hood" of the Constellation program. I tried to explain the role of the transition team and let him know that by not sharing specifics with us directly, we had to turn to other sources. Mike responded by telling me he wanted to speak with the most senior person on the Obama transition team who was looking at NASA. I told him that it was his lucky day: he was talking to her.

Mike was a well-regarded technical leader in the space community, someone I'd known for nearly twenty years. Just a few months earlier, I'd helped him secure Senator Obama's support for an important export issue that allowed NASA to continue their critical strategic relationship with Russia. I remembered Dan Goldin's cooperative relationship with Sally Ride when she was the transition team lead for the incoming Clinton administration, and I'd looked forward to a similarly cooperative experience with Mike.

But the Administrator made it clear in our first meeting that he was uninterested in our efforts. I offered to set up regular touch-base meetings between us, but he dismissed the suggestion. Mike's reaction was disappointing, though the team did our best to set it aside. Over time we learned that he had good reason to be concerned about the questions we were asking about Constellation.

I knew Deputy Administrator Shana Dale, too, and we had a more productive relationship during the transition. Chris Scolese, the Associate Administrator, was the third highest-ranking official at NASA and the most senior civil servant. Chris started working at NASA Headquarters after I left in 2001, so we didn't know each other. He appeared even less interested in working with us than the Administrator, but I hoped his view would change once the new President was sworn into office.

Another notable one-on-one meeting I had during the first days of the transition team was with Robert Cobb. The findings of previous investigations into his questionable actions were public, but the unique nature of the IG position had allowed him to still be retained at that point. "Moose," as he was known, was charming and had a singular message. He told me that he was the kind of IG who liked to work with management and went out of his way to say that he looked forward to working with me. Out of all the people who should have been delivering me that message in NASA, he was one of the few who did. Coming from him, it was somewhat inappropriate.

A few weeks after the election, Mike Griffin's wife Rebecca, along with a former astronaut-turned NASA contractor, circulated a petition and message throughout the aerospace community, asking for me to be removed and for Mike to be retained as Administrator. The controversy garnered mainstream media attention, including an article by Jeffrey Kluger in *Time* magazine, in which he opined that NASA was correct to be nervous about my appointment. He initially referred to me in the article as an HR rep and went on to describe me as a former NASA public affairs officer—I have never been either—who "competed with a boy band singer for a chance to fly on a Russian rocket." Rachel Maddow of MSNBC and other media reported the story in a more accurate way, but I steered clear of commenting publicly, already concerned about failing to adhere to the "No Drama Obama" philosophy. The dust-up increased my visibility to senior transition team officials, but failed to fulfill their intention. Instead of diminishing my chances of being offered a senior position in the incoming administration, it likely helped.

Even before his friends and family began their campaign, it was clear Mike hoped to stay on at the space agency. He had a lot of support on the Hill and in the industry. I wouldn't have been all that opposed to him being retained at least until a new Administrator was confirmed. It wasn't up to me, but I'd learned early that wasn't the incoming President's plan.

As the Agency review team lead, I received an expedited security clearance in advance of the election. John Podesta had been named head of the transition team for Obama, and I happened to be standing in line next to him as we waited to be fingerprinted for our clearances at the FBI. I introduced myself as the NASA transition lead, and we conversed about his long-held interest in the topic. He shared his view that the guy leading NASA at the time seemed "like a real nut job." I asked why he had that impression, and he recalled hearing the Administrator's recent interview about climate change on NPR:

> To assume that it is a problem is to assume that the state of Earth's climate today is the optimal climate, the best climate that we could have or ever have had and that we need to take steps to make sure that it doesn't change. First of all, I don't think it's within the power of human beings to assure that the climate does not change, as millions of years of history have shown. And second of all, I guess I would ask which human beings—where and when—are to be accorded the privilege of deciding that this particular climate that we have right here today, right now, is the best climate for all other human beings. I think that's a rather arrogant position for people to take.

This was not going to work out for Mike. Personnel decisions on Agency leadership were not in the purview of the landing teams, but even if I'd made a recommendation to keep him, there was no chance it would have been accepted. I would have shared this with Mike myself, but we were specifically asked not to get into discussions of tenure with current political leadership. If the incoming President wanted someone to stay, that person would be contacted directly by the personnel team, but otherwise, the across-the-board assumption was that their service ended at noon on January 20. I did what I could to send that signal to Mike through private channels, but he was reportedly still hoping to "get the call" right up until the end and blamed me for not being selected.

Although not directly my responsibility, my goal was to have an Administrator selected and possibly even confirmed before Inauguration Day, since the space community views early attention to NASA as a harbinger of the President's support for space. During the transition I was asked about my own interest in serving in the administration, and for a list of recommended candidates for senior positions. My highest aspiration was to be the NASA Chief of Staff, but taking my dad's advice to seek positions at least one-rung higher than you want, I expressed my interest in being Deputy Administrator. I also submitted a list of seven extremely qualified Administrator candidates, all of whom would have been strong Deputy Administrators.

The head of personnel for the President-elect, Don Gips asked me what I thought of Scott Gration for NASA Administrator in early January.

I told him I didn't think Scott fit the qualifications and likely wasn't interested in the job. I asked Don who made the suggestion and he said it came directly from the President-elect. I changed my answer and said I thought he would be terrific—having long believed that one of the most important characteristics of a successful NASA Administrator was having a close relationship with the President.

Scott Gration had met then Senator Obama on a multicountry trip through Africa several years before. Gration was the son of missionaries and they happened to have discussed NASA, where Scott had been assigned as a White House Fellow decades earlier. During the campaign, Gration was credited with creating the influential group of sixty military generals that endorsed Obama, an effort widely credited with his ability to overcome Hillary Clinton in the primary election.

I had only talked with Scott about NASA once at that point, at the suggestion of John Podesta. Scott was leading one of the transition teams at the Defense Department and immediately understood why I was calling. He chuckled as he relayed his one-year stint at NASA in the 1980s and his brief conversation with the President-elect on the topic when they met in Africa. Scott readily acknowledged that he had no particular insights or recommendations on the best path for the Agency and it seemed clear from our conversation that he didn't expect to be appointed to a NASA leadership role.

Nevertheless, the trade press reported Gration's likely nomination a few days after my conversation with Don. After Senator Nelson made public statements that he didn't think General Gration had the right qualifications, the White House was silent about any such intent. As the political personnel team moved on to look at other candidates, it became clear NASA wouldn't have a nominee before inauguration.

The Agency transition teams were asked to identify acting leadership to serve in the absence of Senate-confirmed candidates and after considering other possibilities, we settled on the Associate Administrator Chris Scolese, still assuming we'd have a permanent Administrator soon. I was ecstatic when the President selected Steve Isakowitz for the position a few weeks after inauguration, then stunned and discouraged when Senator Nelson disapproved and the President demurred.

Government service offers promotions based on time served while also providing job security, which often leads people to spend their entire careers in public service. Chris Scolese, like many of NASA's leaders, had spent nearly his entire career in government, which seemed to ground his views in the past. One of the advantages of entering or returning to government service after "working in the real world" is the outside perspective people bring back to the space agency.

The bursting of the dot-com bubble had slowed the growth of the traditional communications satellite industry in the early 2000s and ended the pursuit of many early investors working to lower launch costs. However, the aerospace landscape had changed considerably since then and a new generation of well-heeled investors had entered the arena.

As 2008 rolled around, the technological advances that had shrunk the size of computers and personal devices were being adopted by new entrants in the commercial space world. Size reduction of satellites drove shorter development times, lower costs, and expanded usership. The success of innovative entrepreneurial companies inspired others to pursue an array of space privatization and commercialization projects. Geopositioning, navigation, timing, and remote sensing from space began as government activities, but they were evolving into large, profitable commercial industries.

In what became a virtuous circle, related disruptive shifts reinforced new developments in the space transportation side of the industry. It helped that the satellite industry finally seemed primed for a boom. With nearly all of the commercial launch market lost to the French, Chinese, and Russians, any US company able to offer reliable, low-cost launches stood to reap a huge return.

As I had told candidate Obama, it struck me as a mistake to have the government continue to design, build, and own rockets. Not only was the private launch market poised for growth, but the industry was already launching non-Shuttle-required payloads. The commercial policies that supported the X-33 and Alt Access programs in the 1990's mapped how the government could offer to stimulate private sector capabilities with greater efficiency. SpaceX and others had their sights set on competing in the arena but needed a way to prove their rocket's reliability, since most customers weren't willing to take the risk of launching on an unproven vehicle.

The 2004 Bush administration space policy, which directed the termination of the Shuttle and a new exploration vision, had also directed NASA to "pursue commercial opportunities for providing transportation and other services supporting the International Space Station and exploration missions beyond low Earth orbit." The Office of Management and Budget had been putting on the order of $100 million in NASA's coffers to initiate a program to do so. In 2004, NASA awarded Kistler Aerospace, one of the private companies that had received initial funding through the Alt Access program, an additional $200 million to help develop their planned reusable launch vehicle. SpaceX protested the award, citing the lack of a competition. When GAO, the government watchdog, told NASA it wasn't going to win the case, NASA withdrew the award and was forced to develop to a new plan.

The program NASA eventually designed in response to the protest (and White House guidance) was called Commercial Orbital Transportation Services (COTS). Like the RLV program ten years earlier, COTS used a partnership arrangement instead of a FAR procurement—specifically known as Space Act Agreements. An additional incentive we'd recommended for the RLV program under Clinton, the concept of anchor tenancy—based on the Kelly Air Mail Act of 1925—was a critical second element of the program. As with so much of this story, it took dedicated and talented personnel to implement the policy that led to its success. One of these early pioneers was Alan Lindenmoyer, who managed the program out of the Johnson Space Center beginning in 2005. Without Alan, and many others who drove creative implementation of these policy ideas, *Escaping Gravity* would be a different story.

NASA awarded SpaceX and Kistler Aerospace—the company NASA had previously selected for funding—COTS development contracts under Space Act Agreement partnership arrangements in 2006. Kistler failed to meet one of its early financial milestones and was replaced by Orbital Sciences Corporation (now Northrop Grumman) in 2007. NASA included an option for partners to offer solutions that could also carry crew, known as COTS-D, but only SpaceX bid on the element.

Their proposal of just over $300 million was not accepted. Mike Griffin made it clear he didn't intend to extend the private sector partnerships to include astronaut transportation, preferring to send money to the Russians until NASA's in-house programs could take over. Congress had signed off on the plan.

Timing worked against us during the 2008 transition in many ways, but we were synced nicely with the American Recovery and Reinvestment Act, also known as the "stimulus bill." Supported by both the outgoing and incoming administrations, the bill offered ways to stimulate the economy during the raging recession. The Agency review teams were asked to propose "shovel-ready" projects that could be funded through the stimulus bill. We canvassed NASA's program offices to find projects that could be immediately accelerated and identified $3 billion worth of programs, including just over $300 million to exercise the COTS-D option for SpaceX to develop a crew version of their cargo capsule—both named Dragon.

The administration ended up requesting $1 billion to fund the Webb telescope, Earth sciences, green aviation, and $150 million to create a new industry competition to transport astronauts to the ISS. The process was streamlined compared to the regular budget, but the funding stimulus package still had to go to the Hill. Chris Scolese—the acting NASA Administrator at the time—worked with senators representing contractors to reprogram more than half of the money to the Constellation program. These funds came partially at the expense of Commercial Crew funding, which ultimately received $90 million. I was disappointed that the administration hadn't agreed to move forward with $300 million for COTS-D or even done much to defend their $150 million request. However, it is possible that an early public battle could have angered Congress enough to zero the request. On this and so much else, we can only speculate.

Another objective I had on the transition team was to have the incoming administration reinstate the National Space Council. The campaign had signed off on my recommendation earlier, so I was hopeful we could make it happen. When I inquired whether Vice President-elect Biden was willing to chair such an entity—the typical structure—the response came back from his office swiftly. It was a hard no. Not wanting to give up, I put forward other concepts for potential council leadership, but in an effort to convey to the public that it was going to streamline government, the President-elect had announced a goal of reducing White House staff by 15 percent. I was told that no new executive councils would be created.

Chris Scolese had avoided me throughout the three-month transition period, only to show up at my door on January 19, 2009—the day before inauguration—to ask me to allow Ron Spoehel, the Bush-appointed Chief Financial Officer (CFO), to be retained. NASA has three Senate-confirmed positions—Administrator, Deputy Administrator, and CFO—so it was an extraordinary request and unfortunate that it was coming to me so late. Hoping to get our relationship on better footing, I told him I'd see what I could do and called the incoming White House personnel office. As expected, they were agitated by the late request. I pressed the issue and the personnel office reluctantly agreed with one caveat. Under no circumstances was the CFO or the acting Administrator to believe the temporary extension would become permanent; the administration intended to fill the position through the regular confirmation process. I relayed this information to Chris, making sure he accepted the temporary nature of the extension. He said he understood and appeared grateful for my willingness to do him the favor.

Charlie Bolden, Chris, and I had lunch together a few months later—shortly after we were nominated. I was shocked to hear Chris say to Charlie, "I think you should keep the CFO." I reminded him, and explained to Charlie, that doing so wasn't an option. As would happen time and again, Chris ignored me and pretended he didn't know what I was talking about. I was genuinely thrown by his ability to blatantly lie to my face on something so significant. I'd been working with White House personnel on filling the CFO position, and by then we had a leading candidate who was being vetted in advance of her nomination.

I tried to outline the reality of the situation to Charlie, but he wanted to interview both "candidates" and make his own selection. After talking to each of them, he decided he wanted to keep Ron. I made another attempt to explain to him why this was not likely to be acceptable to the

White House team, but he didn't want to hear it. As I predicted, senior staff in the Executive Office of the President said no. Charlie seemed angry at being overruled, and the White House team appeared equally upset that he'd even raised the question.

The intensity of leading the transition team in 2008 and 2009, followed by Senator Nelson's unwillingness to support the President's first intended nominees, and months of uncertainty after being asked to serve as deputy in February was strenuous. Still, by the time of our public announcement in late May, I genuinely looked forward to serving under Charlie. We didn't know each other well, but our perspectives, skill sets, and dispositions were different and I believed had the potential to generate positive results.

The son of a South Carolina high school teacher and football coach, Charlie Bolden grew up to serve his country as a marine general, four-time astronaut, and the first African American to lead NASA. Charlie remembers being inspired by the television show *Men of Annapolis*. He was in seventh or eighth grade and "fell in love with the uniform and with the fact that they seemed to get all the good-looking girls." One of only a handful of Black midshipmen at the US Naval Academy, he was elected president of his class and became a marine aviator and test pilot, flying over a hundred sorties into North and South Vietnam, Laos, and Cambodia. After serving as a marine recruiter for a few years, he was accepted to the Naval Test Pilot School at Patuxent River, which is where he was when NASA selected its first three Black men into the astronaut corps. The next astronaut class—announced two years later—included just one African American, Charlie Bolden.

Charlie was a quintessential astronaut candidate. Over one hundred future astronauts first trained at Pax River and over fifty have graduated from the Naval Academy. In 1968 alone—while I was playing with Barbies and dreaming of becoming a stewardess—three future astronauts graduated from Annapolis and went on to train at Pax River. Charlie, Mike Coats, and Bryan O'Connor notched a combined nine Shuttle flights between them before being promoted into senior management positions at NASA. Mike and Bryan still occupied these positions in 2009 and both wielded significant influence over Charlie as a force of opposition to reforms the administration and I tried to advance.

Charlie came to the position as a national hero, having retired from a stellar government career, with seemingly nothing to prove or change. I was fifteen years his junior, and second youngest deputy, with a background in NASA policy, commercial aerospace, and nonprofit space advocacy. I considered public service a privilege and was determined to transition NASA into a more effective agency. The first five years I'd spent at NASA Headquarters in the 1990s had been the most rewarding of my career. Charlie had spent eight months of his forty years of government service assigned to NASA Headquarters in the 1990s. He openly expressed his disdain for people in Washington, and made it clear those years had been his least rewarding.

Charlie described how much he disliked his time in Washington in a 2004 interview, saying in response to a question about his most enjoyable and most challenging memories in his career, "Oh, without a doubt the most challenging in my fourteen years with NASA was getting on the airplane and going back to Washington. And it got harder and harder every time when I'd come home, going back to Washington for that job. Seriously, that was my undoing. I have never hated a job. I hated that job." He explained that "it just wasn't me. Either you like Washington or you don't. It's for power people, so if you go there and you like being with the power people and at least pretending that you're wielding a lot of power, it's a good place to be. If you're not a power person, you don't like it. I didn't like it."

Charlie's friendly and humble manner made him a beloved public figure. Working together affirmed his positive reputation was well earned. But Charlie's pleasant demeanor made it impossible to discern his true intentions. I learned over time that what Charlie said was often incongruent with his beliefs or deeds. I'm sometimes asked if I have any regrets or things I'd do differently as NASA deputy. Not finding a way to develop a more trusting relationship with Charlie tops my list.

Our first dinner alone together came after we were formally nominated, while we were working through our meetings with senators in advance of the full-Senate confirmation hearing. I asked Charlie a question I thought was natural in our situation: "What do you want to do with NASA?"

After thinking about it for a few seconds, he responded, "Oh, I don't know. What about you?" I paused and then laid out what I saw as our biggest challenges and opportunities. Charlie reacted as though he supported what I had described, nodding, and saying that all sounded good. Well, good.

• • •

After completing individual meetings with the twenty-five senators on the Commerce Committee, Charlie Bolden and I were scheduled for our confirmation hearing on July 8, 2009. My mom, sister, and uncle flew in from Michigan and joined my husband and two sons in the hearing room. An overflow room was required to accommodate Charlie's many well-wishers who had taken buses up from South Carolina to show their support. Congressman John Lewis of Georgia—the heroic Freedom Rider—joined a dozen or more senators and members in making stirring speeches on Charlie's behalf. Senator Nelson and Senator Hutchison made long, glowing statements welcoming Charlie. After a staff member sitting behind Senator Hutchison whispered in her ear, she added that she also welcomed me. Senator Debbie Stabenow from Michigan made my nomination official and spoke on my behalf.

I remember everything about the hearing. Chairman Jay Rockefeller, senator from West Virginia, allowed us each to read our prepared formal statements before fielding a handful of questions from the committee. As expected, Charlie got most of the questions and I interjected when called upon. Nothing controversial was raised, and the formalities were over in less than an hour. We retired to Senator Nelson's chambers, where he and others gave speeches about Charlie, occasionally mentioning me to be polite. I was thrilled with the whole process and have said many times that if you ever have to go through Senate confirmation, I recommend going through it on Charlie Bolden's coattails.

We were soon voted out of the committee unanimously, and within a few days, our confirmation was passed on the floor by unanimous consent. A day later—on July 16—we were sworn in together at a low-key ceremony in the waiting area outside our offices at NASA Headquarters. These affairs are sometimes made into a larger ceremony (my predecessor was sworn in by Dick Cheney in the Indian Treaty Room at the White House), but Charlie and I were ready to get to work.

The week of our swearing-in was the 40th anniversary of the first Moon landing, and the crew of Apollo 11 was on hand to participate in several preplanned celebratory events. There was an evening concert at the Kennedy Center, where we sat in the President's box, and a visit to the Oval Office with the astronauts to chat with President Obama. I'd accompanied Neil, Buzz, and Mike to the Oval Office to meet with President Clinton in 1999, but I could never have imagined I'd be back in such an elevated role ten years later. The furniture and decor were different; the protocol and topics of discussion were similar.

The moonwalkers drew quite a bit of attention, and there were a few autograph seekers, even in the West Wing. One member of the National Security staff was especially anxious to show off his relationship with Buzz and pulled the five of us into an impromptu sit-down with the National Security Advisor. General James Jones was a four-star marine general who knew Charlie well, so the conversation was relaxed and friendly. Jones mentioned a policy review his office was undertaking, which was the only substantive part of the discussion.

Charlie and I held an all-hands meeting with the NASA employees the following day.

We hadn't had much time to prepare, but Charlie and I were both comfortable extemporaneous speakers. The stage was set with two raised stools—Regis and Kathy Lee–style. We each made opening remarks and then fielded questions from HQ and remotely from all the Centers. I was careful to let Charlie speak first and longer, which wasn't difficult due to his folksy way of communicating. It was a lengthy, unscripted discussion during which Charlie said a few things I found inappropriate, such as talking about religion. He became emotional and teared up a few times. It was disarming at first, but it became a regular occurance. We had different styles, and his seemed to work for him. Charlie noted we'd met with the President the day before and mentioned talking with the National Security Advisor about their policy review. I flinched at the time, doing my best to move the conversation along, but it was too late.

The National Security Council policy information Charlie disclosed wasn't public. Senior people at the intelligence community supposedly went ballistic. White House and NASA communications staff were told to pull all of Charlie's media interviews indefinitely. I was asked to substitute for anything that couldn't be canceled until we heard otherwise. Charlie was willing to lie low for a while, but it was unrealistic and unreasonable for the new head of a government agency not to be available to the media.

The NASA Administrator and Deputy offices had an adjoining door, so in the first few months it was rarely closed, and we welcomed each other with a friendly hug each morning. Charlie explicitly told me he didn't want to assign me separate areas to lead. He said he considered me an across-the-board deputy. I had an open invitation to attend his meetings, and he assigned about half the senior staff to report to me. Charlie told his leadership team that I would act in his stead when he wasn't available.

Charlie said he'd take the lead on our key congressional relationships. I focused on briefing freshmen members about the value of NASA and enjoyed those meetings immensely. I had positive relationships with members and staff on both sides of the aisle, but there were plenty of other things to do, and I was happy to leave the heavy lifting to him. I kept a few back-channel lines of communication going with appropriations staffers in the House and Senate, but never went up to the Hill unless I was specifically invited.

Human spaceflight is NASA's most expensive and visible activity. And it was the initiative furthest off-track when we arrived. As an astronaut, Charlie was uniquely capable and suited to evaluate the program and provide leadership for a solution. Our transition team findings and stimulus budget request were included in the briefing books and discussed thoroughly in our hearing preparations. The presidentially appointed Committee on the Future of the US Human Spaceflight Program, which had been established to address the administration's concerns about the program and inform our path forward, was more than halfway through their review as we began our service.

The significant technical issues with Ares I and Orion that we had flagged on the transition team were confirmed by the committee. They presented their conclusions to President Obama's science advisor and director of the Office of Science and Technology Policy, Dr. John Holdren, as well as to Charlie, me, and other senior administration officials a month after our confirmation. The briefing was held in the White House complex, with the entire committee in attendance. There were five options outlined in the report, including one to continue with Constellation. That option required an additional $3 to 5 billion annually, and even then, the committee said it was unsustainable and would not get us back to the Moon.

The committee noted that although the program had received every dollar requested through the budget process, it was falling further and further behind. They saw the same fundamental flaws that had plagued the Space Shuttle and Space Station programs. NASA had designed the biggest rocket and capsule possible to fill existing infrastructure, believing it was the best way to garner political support. Full use of existing people and facilities had become the primary goal of the program.

The chair of the committee, former Lockheed Martin CEO Norm Augustine, spoke about Ares I's schedule slips at a public forum at MIT later in the year, and laid out the problem succinctly: "The Ares I program has been underway for four years, and during that four years, it's slipped about five years, and of course that has a huge impact on how it fits with the overall program.... The Ares I had a near-term objective, which was to support the International Space Station. The problem is, with the current budget profile, the International Space Station will be in the South Pacific two years before the Ares I is available."

He added, "There is not any money going into establishing a presence on the Moon today because that money has all had to be sucked forward to work on the Ares I and the Orion. So, we sort of have a dilemma: dressed up for this party and no party.... The question with regard to the Ares I launch vehicle is not so much *can it be built*, as the question is, *should it be built*?"

The committee's support of private industry taking on transport of astronauts to low Earth orbit (LEO) was also well articulated in the report and in our briefing. This was a confirmation of policies dating back to the Clinton administration, so it wasn't all that surprising, but it was good to have verification from people like Norm Augustine, Sally Ride, and the other expert panelists who didn't have an axe to grind.

The committee outlined the various destinations they had considered for astronauts, with a focus on an option they called the "flexible path." They walked through how many times NASA had been given destinations by elected leaders, only to be followed by unrealistic budget estimates and time frames. The strategy was to get enough buy-in to sell the program initially—knowing that it would in actuality cost significantly more time and money. The flexible path option allowed for investment in advanced technologies that would lower the costs and time frames required for any future deep space destination. It seemed totally logical to me.

I left the briefing with a profound sense of satisfaction. These ten brilliant independent experts had confirmed my own concerns about the existing program and had offered several potential paths forward. Although I had been confident in our transition team findings, this was important validation of what we'd learned. I rode back to NASA Headquarters with Charlie and asked what he thought about the report. He said he was impressed. When I asked which of the five options he thought we should pursue, he responded that we could do any of them.

• • •

The annual federal budget request is traditionally transmitted by the administration to Capitol Hill in the first week of February. It begins in the agencies at least six months earlier. The Office of Management and Budget (OMB) leads the process and substance, pushing paper up the chain, breaking logjams between the Executive Office of the President (EOP) and federal agencies. Typically, the agencies draft their budgets based on the previous year's budget and in consultation with OMB subject matter experts. Agencies formally submit their request to OMB in the fall. OMB then reviews the overall budget, passing back specific questions and incorporating each agency's answers into the final submission—usually in December. NASA's process for the 2011 budget cycle proceeded as usual for all programs other than human spaceflight.

The EOP staff who serve as budget examiners work for multiple presidents and political teams, and the NASA OMB team was professional, knowledgeable, and experienced. Paul Shawcross, the leader of that team, knew NASA's budget better than anyone. If you were up to something good, he'd do his job to help you, but if you weren't, he did his job to correct the problem. Any policy ideas or changes we wanted to pursue would have to make it through him and his chain of command to survive. Congress has the power of the purse, but, for the executive branch, the process goes through OMB first.

With the Augustine Committee report in hand in September, it was decision time for NASA's human spaceflight budget request. Memos were circulated within the EOP, and a principals meeting chaired by John Holdren was held in October. Charlie and I were there from NASA alongside the principals or deputies for OMB and the National Economic Council. The briefing materials included their favored option to cancel the Constellation program in its entirety, freeing up funds for technology development, infrastructure revitalization, and a commercial crew initiative. Other options included retaining Orion and accelerating development of a heavy-lift rocket. All of the scenarios presented at the meeting included commercial crew, and none continued funding for Ares I. If Charlie had a preference for any of the options—or wanted to add new ones—this was the time to make his case, but he left the meeting without swinging the bat. It was *strike one* to those paying attention.

We had crystal clear guidance from the administration to start a commercial crew program, but that wasn't what NASA's internal team developed. Without an explicit message from the Administrator to adhere to White House policy, the bureaucracy saw no reason to change course. Instead, Charlie proceeded with a budget that kept all of Constellation without variation. NASA had an opportunity to replace Ares I with a commercial crew program and restructure the rest of Constellation. Several of us suggested to Charlie that it was time to get creative and offer options for a middle ground. My attempts to convey the importance of following the administration's guidance were ignored. I saw the two trains barreling down the same track toward each other and knew if NASA didn't come up with a different program, there would be carnage.

New administrators historically adjust key members of their leadership teams, maximizing effectiveness and assuring alignment with the elected administration if needed. The governmental rules about senior management changes are established to accommodate a waiting period of 120 days after the appointment of a new head of agency. This delay gives existing managers the time to adjust to the policies of recently elected leadership and the opportunity to prove themselves.

As the 120-day waiting period came to a close, I set up a meeting with Charlie to discuss a few key personnel changes that I thought he should consider. Charlie looked at my list and said he didn't think any changes were necessary. I pressed the issue, explaining why I thought it would be helpful to develop a team willing to work in alignment with the administration. It was a delicate topic to raise, but I considered it my responsibility to share my best advice with him directly. After watching Charlie struggle to address conflicting views, I could see that having an aligned, trusted team would be critical to getting NASA back on track. But no staff changes or requests for modified budget plans were directed. Whether it was by Charlie's inaction or his explicit direction, NASA submitted a human spaceflight budget that by all indications would not be accepted by the President.

On November 21, OMB Director Peter Orszag, John Holdren, and head of White House Legislative Affairs Rob Nabors met with Charlie at his request. I'd offered to have the team help him prepare talking points for the meeting, but he didn't want help and didn't share his agenda. Charlie reported afterward that the meeting went fine. White House staff who had attended with their principals relayed that he hadn't asked for anything during the meeting. The crash was upon us and, as far as I could tell, no effort had been made to switch either train to a new track. Peter Orszag and Rob Nabors were particularly uninterested in attending meetings where nothing was being decided, so it was a second strike.

Charlie had one more chance to defend the plan NASA put forward, or recommend modifications, this time to the President himself. The meeting was scheduled to be held in the Oval Office on December 16. Again, we offered assistance to help him prepare, and again he said he didn't need anything; he knew exactly what he wanted to say.

The final decision memo outlining options for the President was being drafted by the White House staff, timed to coincide with Charlie's meeting. I was assured the President wasn't going to make any final decisions until after he met with the Administrator. Four options were outlined in the memo, all with a top-line close to $19 billion and a five-year runout increase of $6 billion. Every option incorporated the guidance given at the earlier principals meeting, with a new program called Commercial Crew replacing Ares I in each scenario. Option one canceled all of Constellation. The budget NASA had submitted to OMB was not one of the options being considered.

The memo acknowledged that none of the scenarios outlined would be politically popular with Congress, especially with members from states with Constellation contractors. It highlighted the fact that option one would require the most significant political capital to earn congressional support. I was eventually shown a copy of the memo with the President's signature and hand-written notes in the margins, but I wasn't given it in advance and neither was Charlie.

Charlie was ebullient when he returned from the Oval Office. He said the meeting had gone well and that the President was engaged on the issues. Charlie shared that they had talked about advanced technologies, like VASIMR—the Variable Specific Impulse Magnetoplasma Rocket. Sensing my nervousness at his mention of the nuclear rocket, Charlie said, "Don't worry; he loved it." When I asked if they had discussed Constellation, he admitted the President had told him he didn't want to do Constellation. Charlie seemed to view it as something he would "keep working on," but it sounded to me like he'd struck out without even swinging the bat.

When I got back to my office, I had a message from White House staff asking why the hell the head of NASA was pitching the President a new rocket without telling him it was nuclear? VASIMR is a concept for an electrothermal thruster, with the potential to reduce the time required for robotic deep-space exploration. It is essentially a nuclear rocket in a very early phase of research, not in development.

White House staff who attended the meeting with their principals described it to me later. They said there had been broad agreement on the Augustine Committee findings, which led Peter Orszag to outline his support for option one. He explained how funding for new technologies would allow for modernization and future programs that could be done in less time and for less money. Charlie had jumped on the "less time" comment, mentioning there was a way to reduce the time to get to Mars from eight months to six weeks. The President hadn't heard of such a program but responded that it was exactly the kind of thing he thought NASA should be doing. Ironically, it seemed Charlie had reinforced the President's inclination to double down on technology development.

Charlie sent an email to those of us who knew about his meeting with the President, saying it had gone well and that he was taking off a few weeks for the holidays. He suggested we do the same. NASA's 2011 budget was the last to be finalized in the federal budget, and we had no one to blame but ourselves. By not following the President's guidance, we had taken ourselves out of the process. The train wreck was now inevitable.

George Whitesides was the first person I'd recruited to join me on the NASA transition team in 2008. George was the executive director of the National Space Society at the time and although I didn't know him well, everyone I respected from NSS and elsewhere told me he was brilliant. They were selling him short. He'd moved into a permanent position at NASA after the transition, and I made sure he interacted with Charlie in the early days, including as the coordinator of our confirmation process. After we were confirmed, I suggested he would make an excellent chief of staff, and the new Administrator agreed. George worked tirelessly to serve Charlie, and at the same time became one of OSTP's and OMB's most trusted and effective NASA leaders.

George Whitesides and I were at the office late about a week after Charlie's meeting with the President when a member of the White

House staff called with the news—for our ears only. President Obama had selected option one, the complete cancellation of Constellation. I was impressed that the President had decided to go with the option that allowed for the most progress, recognizing it was also the most difficult politically.

George's reaction to the news was more subdued. He agreed that option one was the best-case scenario for space development. His concern was that by trying to cancel all of Constellation at once, the push-back could overwhelm progress for our priorities. I'd assumed putting forward a transformational budget signaled that NASA would receive high-level support from the White House. The administration's decision confirmed to me that the President was willing to put his weight behind a bold, innovative plan.

I knew the added weight would be important, since the decision put us on a new track, with another train headed right for us: Congress.

Beth Robinson was our new Senate-confirmed Chief Financial Officer. Recommended to me by Steve Isakowitz when I was on the transition team, Beth had previously served as the most senior career staff member in OMB. She knew the budget process and people involved better than anyone else at NASA. Beth and her team worked with the White House staff at OMB and OSTP over the holidays to respond to questions and start drafting language for our Congressional Justifications (CJs), which the Hill would be expecting to accompany the budget.

NASA had a well-earned reputation for strategically leaking pre-decisional budget information from the White House. Given we had already been obstinate by not following the administration's guidance, trust was at a low point. The President's decisions were finalized in early January, and although George and I knew generally what to expect, we weren't given the details until after the Administrator's briefing.

Charlie was briefed on the budget in late January—the week before its formal release. The magnitude of the decision seemed to hit Charlie hard. He said he felt like he'd been kicked in the gut. I showed him charts from previous principals' meetings, walking him through the earlier guidance and decisions he'd somehow misunderstood. He said he recognized he'd been part of the process but acknowledged he hadn't fully understood his role. I was at least partially culpable. After my early advice and counsel was ignored, I took my cue from him and shared less.

Charlie's demeanor during his budget briefing at the White House had raised alarm bells about his ability to communicate and advocate on behalf of the administration's strategy. Those concerns led to a decision that gave him a more scripted role in the formal budget announcement. None of us liked the plan—least of all me—but the White House didn't want to risk losing the message on the first day the budget was released. In Charlie's defense, the process wasn't that different from how things worked in large cabinet agencies, where the secretary gives a prepared overview and leaves the details to others. The risky alternative was to have the process become the story and distract from the substance of the proposal.

That is indeed what happened.

On the day the federal budget was released to the public, Charlie read a prepared statement on a press call and then said he had to depart for "other meetings." He told the media that policy questions and answers would be fielded by his deputy and Jim Kohlenberger, the OSTP chief of staff. After our Q & A session, Beth briefed the detailed numbers and took questions while I was sent across town, along with Dr. Holdren and other science agency heads, to present the NASA budget at the American Association for the Advancement of Science. I ended the day back at Headquarters hosting a round-robin press event with key members of the media.

The Obama administration's $19 billion request for NASA was a $300 million increase over the previous year, plus the additional $6 billion over the planned five-year runout. It proposed canceling Constellation, which freed up money for an additional Shuttle flight and extension of the ISS from 2015 to at least 2020. New funding was provided for Earth sciences, advanced technology, rocket engine development, infrastructure revitalization, and building on the industry partnership we had begun in the stimulus budget to transport astronauts to the ISS—the program that became known as Commercial Crew.

Given that we proposed terminating contracts worth billions of dollars, the negative response to the budget was not surprising. Congressional committee assignments are chosen according to members' own preferences, which perpetuates self-interested committee leadership. Since NASA hadn't been a part of a larger national agenda for decades, its standard bearers included self-selected senators and representatives with existing contracts and jobs in their districts whose primary interest was often maintaining the status quo.

The federal budget process precludes consultation with Congress or sharing of any information until its formal release, but the aerospace establishment claimed to be shocked that such drastic changes were being proposed without their advance knowledge. No matter how it was announced, canceling contracts worth tens of billions of dollars was never going to be popular, and no amount of advance notice would have changed their minds. Charges that the budget was developed in secret were fueled by parochial interests. The internal NASA team who worked on Constellation had assumed ignoring the White House guidance was enough to get their own way. They were furious and frustrated by their own lack of knowledge about the final decision.

The budget request tracked closely with the public recommendations of the Augustine Committee, which had been formed to review the program and propose future budgets. Their findings had publicly aired untenable problems with Constellation, and the trade press had been speculating about the possibility of major changes in NASA's programs for months. No one could really have expected the administration to break traditional rules regarding holding pre-decisional information to give extra planning time to those they knew would be opposed, but the myth that the *process* was the most significant problem with the budget request prevails.

Congressional hearings were quickly scheduled, and the team helped Charlie prepare. I'd asked for the President, or even the Vice President, to make a handful of calls to the Democrats in leadership positions as a courtesy when the budget came out, but securing votes for health care and other priority programs took precedence. Whether personal calls to the Hill's Democratic leaders would have been enough to stave off their opposition to the plan is unknowable, but it could have provided a context and tamped down claims that the process and decision had somehow not involved the President.

I talked with Senator Nelson and explained the rational for the plan the day it was released and got an initial sense that he might be supportive. As time went on, he appeared to be—somewhat understandably—offended not to have been involved in mapping the strategy. Like many others, buying into the new plan would have required his acknowledgement that the program he had overseen and publicly supported was leading nowhere.

In one particularly uncomfortable one-on-one meeting in his Senate hideaway, he aimed the intensity of his ire in my direction. In response to public comments Elon Musk had made about SpaceX's ability to improve on NASA's existing programs, Bill Nelson shouted at me to "get your boy Elon in line." Given how much the Florida space coast stood to benefit from the proposal's infrastructure investments and development of a robust commercial launch industry, his lack of support was especially disappointing.

Charlie had difficulty messaging both the content of the budget and his role in its formulation, which led even NASA leadership to believe that he had been undermined in the process. NASA senior staff were quick to tell the Hill they hadn't developed the plan that had been submitted to Congress and claimed that the scant technical details signaled a program not quite ready for primetime. They never acknowledged that their own unwillingness to follow White House direction was responsible for taking NASA out of the process or that technical details related to new technology programs are developed when the grants are negotiated, not when they are proposed.

Separating the message from the messenger was challenging, and coming from the number-two spot, what I said fell on deaf ears or worse. Charlie didn't want to be critical of Constellation, concerned about the impact it would have on the workforce. I thought we should be open about how far off track the program was while acknowledging that the problems stemmed from structural decisions and were not a reflection on the workforce. Our different approaches left us without a compelling, consistent message.

Charlie responded to accusations that I was to blame for the budget by saying "many individuals were involved in the budget process."

Instead of focusing on how the proposal would advance a more meaningful purpose for the Agency and drive economic benefits, he said if the panel wanted to blame someone, they could blame him. Whether this had been his intention or not, his response did nothing to quell the impression that this budget proposal was somehow created without his or the President's knowledge.

Anyone who knows anything about driving significant change in large government programs is aware that unanimity among the most senior officials with constituency interests is required—including final sign-off by the commander in chief. The truth was that the program of record was an abomination, and the policies and programs we proposed were neither unexpected nor radical. The tactic to blame a cabal of low-level-functionaries, led by me, for driving an agenda that opposed human spaceflight was cultivated by people with a self-interest in keeping the existing program—including Charlie's own cup boys. Gaslighting prevailed. The message was shaped by the community that was responsible for the problems with the Constellation program and therefore had the most to lose.

The ideas behind the proposal began with me, but ultimately were shared by President Obama, his science advisor, the Director of OMB, the National Economic Council, and the Augustine Committee. I had not spoken to the President directly about his goals for human spaceflight since my time on the transition team. What I heard from everything his staff and Charlie reported from their own meetings led me to believe Obama's views were consistent with what we'd discussed in 2008. The cabal narrative was flattering to me in some ways, since it credited me with outmaneuvering not only the marine general and astronaut Administrator, but also the President of the United States.

Within a month of the budget's release, the press heard that, in response to Charlie's orders, the head of NASA's biggest Center—the Johnson Spaceflight Center in Houston (JSC)—had begun developing an alternative plan to the President's. Mike Coats was one of Charlie's Annapolis classmates-turned-astronaut. He had worked for Lockheed Martin between his Shuttle flights and his NASA management position. Mike was the kind of cup boy who didn't understand why the government would ever trust a company other than the established contractors—like he'd worked for—to get near the human spaceflight program.

Space News reported on March 4, 2010, that Mike had sent out an email saying the Administrator had asked him to gather the other Center directors with responsibility for human spaceflight, along with key leaders at NASA Headquarters, to develop what came to be called the Space Launch System (SLS). The NASA leaders' goal—coordinated with industry and key congressional staff—was to keep Constellation contracts intact. The email referred to the alternative being developed as "Plan B."

It was never clear to me how much of Plan B Charlie directed purposefully, or if it was an outcome of his reticence to say no to his cup boys. Either way, the message was conveyed that he didn't mind if the budget was overturned. NASA put out a statement in an attempt to distance Charlie from Plan B, but the Coats email tying him to it had been copied to eight NASA leaders and was already widely circulated.

In one of his NASA oral history interviews a few years later, Mike Coats discussed his view of Charlie's role in opposing the President's budget and creating his own plan. Coats said that Charlie had found out the White House was going to cancel the Constellation program a few days before the budget was released, saying, "I think probably informed by Lori Garver, but I don't know. From what I've heard, Charlie spent the weekend arguing with the White House staff—chief of staff, probably. 'How about giving me a chance to restructure the Constellation program, instead of just canceling.'" Mike and others solidified a myth around this narrative for their own benefit.

Plan B worked so well that the SLS launch vehicle it created is often credited as the *Senate* Launch System, even though it was created by a combination of self-interested parties in the Agency, industry, and on the Hill. At the end of his tenure, Charlie opened up about his pro-SLS agenda and publicly criticized me for supporting Commercial Crew. In a late 2016 interview, he said that he and I interpreted presidential priorities differently. He blamed me for being on a side with my "political allies in NASA, OMB, and the White House Office of Science and

Technology Policy" favoring Commercial Crew over SLS/Orion, while Congress was far more favorable to SLS/Orion, as were career senior leaders at NASA associated with human spaceflight.

After publicly expressing his support of Commercial Crew for more than seven years, his new narrative castigated me for backing the program. His charge was that, unlike him, I had "interpreted presidential priorities" in alignment with the Executive Office of the President instead of Congress. Guilty as charged. By finally acknowledging he was on the other side, Charlie revealed that the difference in policy priorities hadn't just been between him and me; they had been between himself and his commander in chief.

6.

HEAVY LIFT

THE GOAL OF COMMERCIAL CREW—SIMILAR TO COMMERCIAL CARGO—WAS to drive down the cost of transportation to and from space. Reducing launch costs would help nearly every aspect of NASA, including Earth and space science missions where launch costs sometimes exceeded the cost of the spacecraft. The new program was mostly being criticized because it competed against Ares and Orion, which were feeding off of traditional cost-plus contracts to the tune of $3 to 4 billion a year whether they flew or not. The benefits of Commercial Crew, the source of criticism against it, and the exorbitant costs of the status quo were parts of the messaging I thought needed to be emphasized beyond the space community. White House legislative affairs officials were focused on other priority initiatives and appeared reluctant to engage on NASA's issues.

I tried to communicate with NASA's employees, contractors, the media, and aerospace associations about the strengths and long-term benefits of our plan. It seemed to have little positive effect. Without top cover from the White House, and being muzzled on why Constellation was being canceled, we dug a very big hole for ourselves. Constellation's industry partners and congressional stakeholders smelled blood in the water and joined forces against us. In hearing after hearing, as the administration and the plan were berated, we never offered much of a compelling defense.

By early April 2010, the negative headlines and calls from angry senators had finally gotten the attention of the White House. In an effort to court Senator Nelson, it was decided the President would visit the Kennedy Space Center in Florida and show his support for NASA. To break the logjam, the administration expressed a willingness to offer concessions if they could secure our top priorities of Commercial Crew, the Webb Telescope, technology, and Earth sciences. The speech was designed to extend an olive branch to the space community, while personally explaining and embracing the plan the budget proposed.

The increased cost, risk, and schedule delays of the Ares I rocket made it the most vulnerable element of the Constellation program. The contract had been sole-sourced to ATK Space Systems—a Utah-based contractor supplying virtually the same solid rocket motors for the Shuttle. Ares I was known as the "Scotty Rocket" after former astronaut Scott Horowitz, who had designed it when he returned to NASA after working for ATK—an arrangement that any impartial Inspector General would have likely investigated. The Ares I was supposed to serve as a precursor to a larger rocket that NASA, industry, and the Hill really wanted to build, known as the Ares V.

In an attempt to attract some support from the industry, the administration's first compromise offer was to retain a streamlined Orion capsule without the Ares I rocket. Lockheed Martin had at least competed to win the Orion contract, and it was early enough in the program that it could have been modified to launch on an existing Evolved Expendable Launch Vehicle and used as a lifeboat for astronauts on the Space Station. Canceling the Ares I rocket would provide the funding needed for Commercial Crew and keep the government from competing directly with the private sector.

The second compromise the White House was willing to make was to identify the next deep-space destination for our astronauts. Astronaut destinations are often confused with goals, and a primary criticism of our proposal leveled by the status quo was that it didn't set a time frame or "goal" for astronauts to go beyond low Earth orbit. President Bush (43) had proclaimed that astronauts would return to the Moon by 2020, but like his father's 1989 proclamation, the son's didn't lead to a NASA program that could deliver on his rhetoric.

Since Apollo, NASA's human spaceflight strategy had been driven by the *what*, *when*, and *how*, instead of the *why*. Enabling NASA's readiness to deliver on a range of missions that could address national objectives for less time and money was a worthy goal, but it didn't work if the *why* was creating and securing jobs in congressional districts. The flexible path we proposed would invest in technologies to reduce the amount of money and time required for all future robotic and human exploration beyond low Earth orbit. The plan was designed to allow the private sector to develop their own capabilities without government competition and give flexibility to a future president to select a destination that took advantage of new innovative technologies and aligned with pressing national and global considerations. The 44th President wanted to make real progress instead of empty proclamations, but that didn't appear to be the goal of those battling on the other side.

The President would only agree to indicate the next intended place astronauts would go if a meaningful destination could be achieved realistically within our budget profile by 2025. The Moon required an expensive lander. Mars cost even more and took longer. Going to an arbitrary point in deep space was an option, but one we dismissed as insignificantly interesting or worthwhile. An asteroid mission was attainable because it eliminated the need for an expensive lander (due to an asteroid's low gravity environment), it utilized space technology systems that were already under development, and it could be supported by Orion.

Asteroids are critically important objects to study for scientific research, possibly carrying the seeds of life; for long-term space development, they provide mining resources that could be used to build stations or starships in the future; and, most importantly, they could impact Earth and could potentially wipe out humanity. This clear rationale makes asteroids compelling to the public, an interest Hollywood has regularly tapped. Even the process of selecting candidates to visit would be valuable, since it would require improvement of our asteroid detection and characterization capability. Finally, the mission would provide an analogue to study the human body's reaction to extended time in deep space—one of the biggest unknown obstacles to expanding human presence beyond Earth's orbit—including to Mars.

The President had time for only one tour and photo-op during his Kennedy Space Center visit, and those of us working on the trip ultimately recommended he go to the SpaceX launchpad. Charlie hadn't been involved in the speech or the planning and didn't seem thrilled when he heard Obama wasn't touring the Shuttle facility. I defended the decision, reiterating that the message of the visit was on the future and not the past, and he didn't pursue a change in venue.

On April 15, 2010, at the Kennedy Space Center, President Obama announced the United States would send astronauts to an asteroid by 2025 and would restore the Orion capsule, which was to be streamlined and used as a lifeboat for the astronauts on the ISS. These two significant adjustments to the initial proposal were sincere attempts to find a compromise that would allow us to move forward with the other important elements of the NASA plan.

When President Obama finished his remarks, he stepped off the stage and walked over to shake my hand. The crowd was still cheering, and I was surrounded by well-wishers that included Neil deGrasse Tyson, Bill Nye, and astronaut Ed Lu. Smiling and patting me on the shoulder, Obama asked me, "Do you think this will help?" I responded honestly that if it doesn't, nothing would.

Unfortunately, I was right. Nothing helped.

Neither offer placated the opposition. Criticism for not having a destination was replaced by criticism that it wasn't the destination they wanted. The existing constituency studying asteroids was too small to influence the budget, and the aerospace industrial base had made a pact among themselves and the Hill to fight cancellation of any Constellation contracts, so launching Orion on an existing launch vehicle was a nonstarter.

Even though the White House legislative staff said they had been working with the Democrats on the Hill and people at Lockheed to assure they were onboard with the adjustments to the plan, by the time I left Florida the opposite was clear to me. I'd talked with Senator Nelson, the Lockheed Martin CEO, and Charlie, and none of them thought what the President said changed anything. Senator Nelson still planned to lobby for a heavy-lift vehicle, Bob Stevens (the CEO of Lockheed Martin) didn't intend to change Orion, and Charlie's takeaway was that we were going to send astronauts to Mars.

Rob Nabors was a White House liaison to the Hill and one of our key contacts in the administration. Every time I saw Rob, he asked me to get Senator Nelson to stop calling him. My consistent reply was that we had been clear that the President's decision was not likely to be supported by the Southern congressional delegation, and nothing I could say was going to change that fact. Only a call from the President, or possibly the Vice President, had the chance to convince Senator Nelson to support the plan. He was never able to make those calls happen, and Senator Nelson kept calling him.

As we left the Cape, Rob was as upset as I'd ever seen him. He told me that for my own sanity, I should get out of the aerospace field. He said, "None of these people care about the actual space program; they are vipers and NASA is a viper pit." I'd been inoculated through years of absorbing venom, but I understood his frustration.

The few days of attention the White House staff gave to the modified NASA plan were centered almost entirely on the President's trip to the Kennedy Space Center. The goal of the communications and advance teams was to deliver our message and drive positive press coverage without taking too much of the President's time. George, OSTP staff, and I did our best to run alongside the moving train and make the most of the opportunity to demonstrate presidential leadership to the space community. The Orion compromise, asteroid destination, and SpaceX visit had been done on our recommendation, with the support of the President's senior advisers, but once Air Force One was wheels up from the Shuttle landing strip, the ball was back in our court to carry the message forward.

In addition to the Apollo astronauts who testified against Obama's budget, active astronauts were openly derisive of the Commercial Crew concept and the recently announced asteroid destination. I made a point to meet with the current corps during one of my visits to JSC. I was disappointed by the low level of interest. The astronauts who did agree to meet appeared openly hostile toward the administration's plan and to me personally.

Their own friends and former astronaut colleagues working as Constellation contractors had misrepresented the program's progress and had falsely conveyed that they were on track to take them to the Moon and Mars. Some of the astronauts undoubtedly had hopes of being hired by the same companies in the future or walking on the Moon themselves. Astronauts also enjoyed their rotations in Star City—getting to know their cosmonaut comrades in the fraternity-like culture—and were not anxious to trade that for being launched on something new, built by the likes of Elon Musk. Not wanting to blame the NASA Administrator, their astronaut brother, it seemed easier to blame me for threatening their dreams, livelihoods, and possibly their lives.

NASA leadership, like most everyone else, is typically deferential toward astronauts, but I'd had enough of being scorned for telling the truth and actually trying to drive to a better future. I reminded them that the true stakeholder of America's space program was the US taxpayer, not them, or any of us who were government employees. I explained that the President's plan to transform the Agency was designed to lead to sustained advancement in space, not to replicate stops and starts that had led us nowhere. My rant likely fed the caricature that had been created of me, but over time this reality started to sink in.

My focus on getting the astronauts to support the Commercial Crew concept had started with the release of the Augustine report the year before. In October of 2009, I had traveled to Russia to meet with the Russian Space Agency and to welcome home astronaut Michael Barratt, who was returning from six months on the ISS on a Soyuz. The number of astronauts stationed in Star City had grown since I was there in my Astromom days, a result of the Soyuz being their only ticket to space after the Columbia accident.

The original "cottage" the United States had built to house the first astronaut to fly on a Soyuz in 1999—Navy Seal Bill Shepherd—was the hub of the astronauts' social activity. The bar in his basement, known as Shep's Bar, was legendary. I was pleased to be invited to join several astronauts there for a celebration of the successful landing. They included Michael Foale, Tracy Caldwell Dyson, Sunita Williams, Mark Polansky, and Michael López-Alegría. My reputation as an advocate of replacing the Ares I launch vehicle with a privately developed launch system was well-known, so my agenda was obvious to them.

The astronauts were upfront about their own agenda when I arrived, telling me that the last Deputy Administrator to visit Shep's Bar got so drunk they had to carry him out. They wanted another notch on that belt. I'm not a big drinker in any circumstance, but I was even less inclined to partake given that my oldest son, Wes—a senior in high school who had studied Russian—had accompanied me on my trip (at my personal expense). When Wes discovered the bar had a piano, he engaged himself by taking song requests as Tracy and Suni joined him on the piano bench, singing "Rocket Man" and sneaking him margaritas.

Hours of debate over the value of privately developed space transportation, many games of Liars Dice, and much margarita drinking ensued. Judging from how I felt during my meetings the next day, I was lucky to have walked out under my own power.

I'd lost my share of rounds of Liars Dice, but a few years later it was clear I'd won the argument. Mike López-Alegría went on to become president of the Commercial Spaceflight Federation; Mark Polansky became a consultant on commercial space issues; and Suni Williams was named to the first operational crew on Boeing's commercial crew vehicle, called Starliner. We could have used their support earlier, but they were important hard-fought wins. Having astronauts on the team was critical, and attempts to get more of them onboard started paying off.

Recently returned astronaut crews regularly brought Charlie and me signed montages with flown flags from their missions. After the formal presentations, we all sat down to discuss the highlights of their flights. When it became obvious these gatherings were mostly social calls, I requested separate meetings to explain our future path for human spaceflight. Not all the astronauts appreciated my inserting myself in their busy post-flight schedules, but several have since told me that our discussions were definitive to their thinking and their later careers.

In July 2010, twenty-four former astronauts signed a letter to the Hill supporting private sector development of the system to replace the Space Shuttle. The effort was coordinated by the Commercial Spaceflight Federation, whose staff operated as my "kitchen cabinet." Their endorsement proved to be an important milestone and helped gain support among legislators.

• • •

Next to astronauts, propulsion engineers or "rocket scientists" (their more common moniker) pull the most weight at NASA. Just as hammers look for nails, rocket scientists look for opportunities to build rockets. Similar to how destinations for astronauts were viewed, building a big rocket—especially one designed to carry astronauts—is seen as an end goal in itself. In my view, having the government build a big rocket was not only expensive and unnecessary, it wrongfully competed with our own private sector.

There had never been a healthy debate over whether the public should be paying NASA to build and operate its own big rocket after the Space Shuttle, so there was no agreement on the fundamental purpose for the system. The goal became feeding the giant self-licking ice cream cone. Elon Musk and Jeff Bezos told me they could develop the capability sooner if NASA served as an anchor tenant, as it was on Commercial Crew and Cargo. Instead of building a direct competitor to US industry—which isn't actually allowed in government policy—the billions of taxpayer dollars saved could have been invested in more valuable missions.

Sacrificing Ares I to start developing a bigger, even more expensive rocket was a brilliant stroke by our opponents that paid off for them. Under the threat of cancellation, the Senate staff, Constellation contractors' lobbyists, industry associations, and the NASA Plan B team drafted legislation aimed to force the country into building a big rocket immediately and to do it with existing Shuttle and Constellation contracts.

Charlie and I happened to be in the back seat of his car together when Rob Nabors called to tell us that the administration made a deal that added building a big rocket to its growing list of compromises. In my view, it was worse than capitulation. Jamming a new big rocket into the budget five years earlier than planned, along with Orion and our own priorities was extremely unrealistic. Charlie seemed relieved, but it didn't appear to me that he'd been involved in the negotiations. As usual, the White House asked me to make the agreement "look like a win." I didn't know if that was possible, but I assured them I would do my best.

George Whitesides, my partner throughout the transition and for the first year of my time as deputy, had called it quits that spring to accept an offer from Sir Richard Branson to be CEO of Virgin Galactic. I missed his strategic thinking and called to get advice. I let George know we'd made a deal that basically reinstated Constellation and asked how we could keep the space pirates onboard. He responded by saying he thought they would continue to support the plan if I was the one who asked. Tears flowed as I thought of how my efforts had fallen short. Even the President's endorsement couldn't abrogate dissent from the NASA Administrator.

We'd been on the call for about ten minutes when I heard a little rustling on George's end and I realized I'd called him out of the blue without asking if it was a good time to talk. Only then did he admit that he was in the hospital with their first child, who had been born a few hours earlier. I congratulated him and apologized profusely, wondering why he had even picked up his cell phone. He replied that it was fine; Loretta and baby George had been sleeping when I called and were just waking up. The awe and wonder of having a new life in the world—a combination of two of the best people I knew on the planet—put my problems in perspective and I steadied myself for the challenge ahead.

White House Communications set up a dozen media interviews with the guidance to spin the deal to show that we had gotten the core of what we wanted. The administration's press officer assigned to NASA, Nick Shapiro, was a classic old-style press guy, and he started each call with the reporters off the record, throwing in some curse words as he laid out the situation. Then he'd introduce me to answer questions on the record. We were a good team, and the hours we spent on the phone that afternoon paid off. The Hill did their own spinning, saying they had won—either a sign of a solid compromise or too little definition in the agreement. The space pirates did what George had predicted and rallied to support the compromise.

I feared the deal would slow NASA's progress in our priority programs in human spaceflight, technology, and science. The Hill would have the last word on funding levels, and the House hadn't even been consulted on the proposal. There were no guarantees they'd deliver the support they promised for Commercial Crew or the technology and science programs. There wasn't enough money in NASA's budget to cover all the priorities we'd signed on to support, and the legislation dictated a number of impossible mandates related to the development of a heavy-lift launch vehicle. Without the Administrator expressing similar concerns, I couldn't get the administration to focus on the issue. Once the House agreed to the Senate language, President Obama held his nose and signed the bill in October.

The legislation allowed us to move forward with our nascent Commercial Crew program. Given that the opposition included most of NASA's leadership, this was no small achievement.

The deal set a competition in motion between the government and the private sector, the dinosaurs and the furry mammals. The mammals would have to compete among themselves and do so while surviving on the dinosaur's scraps. I knew they would eventually be successful, but I hoped it wouldn't be on evolutionary time scales.

Once the bill was signed, the parties each went back to our corners to get stitched up and strategize for the next round—the all-important battle over which rocket to build. We'd agreed to build a government owned and operated system, but a few of us—including many in the White House—still wanted to assure that whatever was built would be designed to be more sustainable than Constellation had been.

The legislation directed NASA to design and build a rocket that could initially launch 70 megatons (MT) and evolve to launch 130 MT, as well as astronauts, to the Space Station and to lunar orbit. The 70 MT version with Orion was to launch by the end of 2016 for $11.5 billion. My response was that you could legislate that the sky was purple, but that didn't make it so. The bill included a requirement to use existing contracts "to the greatest extent practicable." Practicable is a term of art meaning "that it can be done," and what they were asking NASA to do most certainly could not be done. Last I checked while writing this book, the sky was not purple, and the rocket has yet to fly. Congress, the contractors, and most of NASA appeared undeterred by the reality of the situation.

Legislating impossibilities in NASA authorization language wasn't a new phenomenon. Bills passed in the previous decades had authorized billions of dollars more than NASA had requested—or that appropriators had approved. The bills became wish lists of programs and studies that were easily ignored. The National Space Society had worked to get authorization language in 1988 that required NASA to report every two years on what it was doing to support space settlement, but no reports were ever submitted, and no one cared on the Hill or anywhere else.

The 2010 Authorization Bill garnered outsized importance because the authorizers strategically aligned themselves with the appropriators. It was an astute power move; the appropriators were unlikely to complete their bills on time. The only thing required to buy in was a bit of magical thinking—believing that utilizing existing parts of expensive programs and piecing them together would lead to bargain basement prices.

The NASA Space Act was amended in 1985 to direct the Agency "to seek and encourage, to the maximum extent possible, the fullest commercial use of space." Congress ignored this principle and disregarded the increased cost and schedule that would result from their direction to extend existing contracts. Utilizing existing contracts didn't fit the definition of "practicable," which in my view required NASA to utilize its legal latitude to implement the legislation within the laws of physics.

The bill directed NASA to determine the design of the rocket and report it to them within ninety days. It was our last opportunity to assure the government's investment in a heavy-lift capability would lead to a lower-cost, sustainable vehicle. Beth and I worked with OMB to commission an independent cost assessment designed to inform the ninety-day report, believing that honesty and truth could still carry the day. We were wrong.

Instead of focusing on the objective of launching heavy payloads in the least amount of time and for the least amount of money, those at NASA designing the rocket and writing the report to the Hill selectively followed the three words in the bill needed to get the answer the industry wanted—"utilize existing contracts."

Instead of working toward an end-state goal of a reliable and sustainable rocket, they went about designing the system like they were playing Twister—left foot in Utah for the solid rocket boosters, right hand Alabama for the core stage, left hand in Mississippi for the tank and engine testing.... You won the game just by staying upright; no need to achieve more.

Given that the single US company that produces large, solid rocket motors is based in Utah, their delegation was particularly concerned about any attempts to reconsider ATK's sole-source contract that they had on both Shuttle and Constellation. In their minds, the legislative language *required* extending existing contracts, and they believed not doing so would be "breaking the law." In my view, considering alternatives that could lead to developing the most efficient and effective rocket was the only way to achieve a "practicable" solution.

The Utah delegation requested an audience with Charlie and me to discuss the issue. The seven men and I sat around a large conference table in the Senate chambers, as we were chastised for considering any architecture other than one that incorporated solid rocket motors. Their message was directed at me, but they asked both Charlie and me to confirm that we would each "follow the law" and use solid rocket motors. I affirmed I would follow the law but made no promises beyond that statement.

Senator Orrin Hatch pulled me aside before we left the meeting, shaking his finger in my face. "I know you are the problem here," he said, "and I am watching, so you better pay attention." The following morning, the Utah papers were filled with quotations from the members of the delegation, crowing that they had scolded the NASA leadership and received assurances that we would "use solids and follow the law"—never acknowledging that those were not the same things. Senator Hatch's media line claimed the meeting was held because "NASA has signaled an interest recently in possibly circumventing the law." Several of the reports on the meeting highlighted that it was in fact legal to develop a rocket without solid rocket motors if it were more "practicable," while others got it mostly right, transcribing it incorrectly as "practical."

Attempting to meet all the impractical and impracticable demands in the legislation took NASA a lot long longer than ninety days to address, leading to rumors that Beth and I were personally holding up the report. The charges were entirely bogus. Nevertheless, in July 2011, with no final report in hand, the Senate sent subpoenas for emails related to the Space Launch System for me, Beth, and Charlie.

A lawyer from the White House general counsel's office called to ask if I needed the administration to exercise executive privilege, and I said no, none of my correspondence would be problematic for the President. The only meaningful addition I made to the report was securing a future competition for a reusable booster. (But even when Blue Origin put in an extremely competitive bid for a partnership agreement, NASA stuck with its ATK contract.)

After the flurry of headlines about having been served subpoenas, the press was silent when no evidence of wrongdoing was discovered. I occasionally see references to being subpoenaed by the Hill on discussion groups as "proof" that I did something illegal or unethical, not recognizing that any such a finding would have been made extremely public. My experience is a reminder that when subpoenas are served, they don't prove that those who've been served acted inappropriately; they may even prove the opposite.

During the infighting over what NASA's next big rocket should look like, NASA's most recent big rocket made her final voyage. Mission STS-135—the final Space Shuttle flight—was carried out by the Atlantis and her crew of four astronauts that summer. Flying out the last nine Space Shuttle launches safely was a responsibility I took seriously, and my respect for the teams who made each mission happen is immense. After all, history is replete with mistakes that occur during the end of long-running programs. Success required keeping attention and focus through the last moment, and the NASA-industry team delivered.

NASA's plan for how to dispose of the Shuttles after their retirement had not yet been finalized when we on the transition team began our work, but the intent was to display them at the Agency's Centers with direct ties to their development. The estimated cost to prepare each orbiter to be "museum ready" was $20 million, not including the transportation costs. The incoming political team, primarily George and I, proposed holding a competition that would allow museums to offset the costs to the taxpayer. The most heavily weighted criteria we developed for selection was accessibility to the largest potential visitor population. Although NASA made private deals to award the Smithsonian and the Kennedy Space Center their Shuttles for free, museums in New York City and Los Angeles reimbursed the government $20 million for the honor to display the public treasures.

The process of delivering the orbiters to the museums was logistically challenging and personally heart wrenching and Charlie had asked me to take the lead. Wes had gone off to college the year before and Mitch was soon to follow, so my message compared the emotions with those experienced when children leave the nest. For me, it was a mix of sadness, relief, pride, and joy. A big part of me was sad to see the Shuttles go, but I also knew we hadn't raised them to be around forever. Just as David's and my children leaving home made way for new adventures on our own, retiring the Shuttles opened up new possibilities for NASA.

I did my best to honor the program's legacy, while looking to the future. My farewell messages focused on the missions of the Shuttle, instead of the facts and figures about its thrust-to-weight ratio. I highlighted the benefits our nation and the world received from the communications satellite, Earth sciences, and national security launches in her early years, and the discoveries revealed by planetary missions, five Hubble-Space-Telescope-servicing missions, and the Space Station, built with more than thirty Shuttle launches.

In my view, a few of the most important legacies of the program were that the first American women and minorities traveled to space on the Shuttles and that international partnerships had played an important role in many of its missions. In these ways, the Shuttle had been "transformative." It is impossible to put a price tag on what the world has gained from our working peacefully with the Russians in space over the past twenty-five years.

I am critical of design decisions made to the Shuttle based on reasons other than operational cost consciousness and alleviating the risk of space transportation, but that in no way reflects on the program's ultimate achievements or the tens of thousands of people who worked to make those achievements possible. Just the opposite. My angst stems from a belief that the workforce and the country deserved better. Recognizing that private companies are incentivezed to focus on returns to their shareholders and Congress owes allegiance to their constituents, the administration's actions—of which NASA is a part—deserve a fair share of the criticism.

NASA and its stakeholders typically blame OMB for its shortcomings—believing more money would solve all ills. My grad school program distributed copies of the original memos from OMB telling NASA their budget for the Shuttle wouldn't be what the Agency had requested, as proof of a smoking gun. What academics miss is that NASA has the responsibility to propose activities that advance national objectives supported by the administrations they serve; ultimately it is the Agency's responsibility to design programs in alignment with the administration's budget. The real smoking gun is that instead of proposing rational budgets, NASA and industry leadership publicly sign on to deliver programs they favor, for unrealistic timeframes and budgets, fully aware that they can't deliver.

I am extremely proud of having been a part of the Obama administration. Out of all the presidents in my lifetime, there is no other I'd rather have served. That didn't keep me from recognizing our shortcomings. Presidents are elected to have the best interests of the entire nation at the forefront. It is their responsibility to select the best people to carry out their priorities and hold them accountable. After proposing a plan in the best interest of the nation, I was disappointed by the amount of political resources expended to gain its acceptance.

The President selected and expressed his strong support for a transformative NASA agenda. As all presidents must do, he counted on others to develop and advocate for the program he put forward. When the inevitable pushback came, he allowed his team to concede to those with parochial interests, without expending political capital. The proposal was designed to stimulate progress, and, with aligned NASA leadership, I believe could have been implemented successfully.

The administration took the high ground to initiate the independent cost analysis, which indicated that the congressionally dictated cost and schedule could not be met. But the White House and NASA Administrator allowed the senators to cherry-pick the language in the assessment to make it appear that the results were an endorsement of the plan. Senators Nelson and Hutchison charged that public release of the results from the independent analysis equated to sabotage, coercing the administration into a final meeting to settle the issue on their terms in September 2011.

The delegation from the administration sent to the final meeting was led by OMB head Jack Lew, and included Charlie, Rob Nabors, and me. We sat obediently while being berated for proposing a "controversial" program without consulting the senators or their staff. The two senators, who were mostly protecting their states' pork barrel projects, grandstanded and claimed credit for "repairing damage" we had done. They got little pushback from our side. Instead of defending the value proposition of our program or reminding them that the status quo programs they pursued were known to be inefficient and ineffective, we folded. We had a royal flush, and they had a pair of twos, but we walked away from the table.

The Senate staff had what they wanted and scheduled a press conference to announce the Space Launch System design in the Senate the following day. Senator Nelson opened the September 14 event and described what he called the "monster rocket," referring to it as if it were already built. Senators Nelson and Hutchison had released a press statement only a week before claiming "the administration was seeking to undermine the manned space program," but the day's announcement was billed as a coming together of the White House and the Hill. Easels displayed large posters that Senator Nelson turned over to reveal artists' renderings of the new design, marked to resemble the Saturn V rocket that took us to the Moon. The detailed graphics made them look like photographs. The Florida Senator stated that "in the bosom of America there is a yearning for us to explore," adding that NASA was tasked to "explore the heavens." Charlie was invited to speak briefly, but only Senators Nelson and Hutchison took questions.

I was standing to the side of the room and had to steady myself against the wall, hardly believing what I was seeing. The ceremony had obviously been planned long before it was approved by the administration at the previous day's meeting. NASA staff from the program offices, Centers, legislative affairs, general counsel, and even public affairs had been working against us in secret. I thought about how many people in the room and across the country were ecstatic with the announcement, unaware that their leadership was lying to them about what was achievable. Thousands of people would spend their next decade working on systems that weren't sustainable over the long term. I felt like I'd failed the workforce and the country.

The announcement was heralded as a victory in most corridors of the aerospace community, with only a few exceptions. The Space Frontier Foundation was one of the rebel groups that had spun out of the National Space Society. Foundation president Rick Tumlinson's comments were on target:

> The amazing possibilities offered by engaging commercial space to lower costs and develop a sustainable long-term infrastructure to support NASA space exploration, settlement, and a new space industry have been trumped by the greed, parochialism, and lack of vision of a few congressional pork barrelers intent once again on building a government super rocket. We've been to this party before; it was a bust then, and it will be this time as well.

The organization's chairman, Bob Werb, added, "Senator Nelson called the SLS a monster rocket and he's right. Its budgetary footprints will stamp out all the missions it is supposed to carry, kill our astronaut program, and destroy science and technology projects throughout NASA."

Not surprisingly, all of the aerospace companies and industry groups backed the SLS decision. The Aerospace Industry Association statement read, "Even as our economy struggles to recover from recession, the plan is a ray of hope that America's belief in a better future endures and America's continued leadership in space exploration can be preserved."

ATK even embraced the planned booster competition, saying they were "well positioned to compete for the final design."

One member of Congress spoke out in opposition to the plan. Congressman Dana Rohrabacher, a Republican from California, said in a statement, "There's nothing new or innovative in this approach, especially its astronomical price tag, and that's the real tragedy." He said he feared that budget pressures would bring this program to an end in much the same way the Saturn V was terminated because of budget cuts to Apollo. "Nostalgic rocketry is not how great nations invent their future," he concluded.

Once the administration signed on, it was our job to do what was possible to give the NASA team its best chance for success. The administration held up its commitment and requested over $3 billion for the program every year, while Congress made deep cuts to much smaller Commercial Crew and technology requests, funneling what they cut to SLS and Orion's much larger budgets. The dinosaurs ultimately received ten times more funding than each Commercial Crew competitor. My objective was to protect the mammals for the first few winters, long enough for them to evolve and create a more sustainable space program.

President Obama's vision for NASA partially was impacted by the same wishful thinking that affected some of his plans in other areas. That is, he believed in his ability to win over the other side. By trusting that all parties had the best interest of the space program in mind, we were often rolled under the bus. The aerospace community's negative reaction to his NASA proposal disincentivized increased engagement. His speech in April 2010 was the only address solely dedicated to NASA during his presidency.

The most often recited words from the President's speech are "been there, done that—right, Buzz?" in reference to the Moon. The line wasn't in the prepared speech that I'd seen the night before. Buzz was on stage with him and the remark appeared extemporaneous.

The overwhelming reason we hadn't adopted the "planned" Moon landing was the acceptance of reality: there was not, nor had there ever been, money for a lunar lander or heavy-lift rocket in the five-year budget runout. Clearly some administrations feel less constrained by budget realities than others, but we thought it was best to tell the truth.

The substantive message from Obama's Florida speech is worth remembering:

I am 100 percent committed to the mission of NASA and its future.... We will ramp up robotic exploration of the solar system, including a probe of the Sun's atmosphere; new scouting missions to Mars and other destinations; and an advanced telescope to follow Hubble, allowing us to peer deeper into the universe than ever before.... We will increase Earth-based observation to improve our understanding of our climate and our world.... And we will extend the life of the International Space Station likely by more than five years...

We will invest more than $3 billion to conduct research on an advanced "heavy-lift rocket"—a vehicle to efficiently send into orbit the crew capsules, propulsion systems, and large quantities of supplies needed to reach deep space. In developing this new vehicle, we will not only look at revising or modifying older models; we want to look at new designs, new materials, new technologies that will transform not just where we can go but what we can do when we get there...

We will increase investment—right away—in other groundbreaking technologies that will allow astronauts to reach space sooner and more often, to travel farther and faster for less cost, and to live and work in space for longer periods of time more safely. That means tackling major scientific and technological challenges. How do we shield astronauts from radiation on longer missions? How do we harness resources on distant worlds?

Early in the next decade, a set of crewed flights will test and prove the systems required for exploration beyond low Earth orbit. And by 2025, we expect new spacecraft designed for long journeys to allow us to begin the first-ever crewed missions beyond the Moon into deep space. So, we'll start—we'll start by sending astronauts to an asteroid for the first time in history. By the mid-2030s, I believe we can send humans to orbit Mars and return them safely to Earth. And a landing on Mars will follow. And I expect to be around to see it.

We will partner with industry. We will invest in cutting-edge research and technology. We will set far-reaching milestones and provide the resources to reach those milestones.... For pennies on the dollar, the space program has improved our lives, advanced our society, strengthened our economy, and inspired generations of Americans. And I have no doubt that NASA can continue to fulfill this role.

President Obama's 700-page memoir, *A Promised Land*, devotes less than a page to NASA. The book notes that as he boarded *Marine One* with his family to head to the Cape for a planned Shuttle launch on April 29, 2011, he gave the go-ahead for the Abbottabad mission—the Navy SEALs' Pakistani raid that led to the death of Osama bin Laden. This confluence of events is a poignant reminder of the many demands on a president's time and how many critical decisions the POTUS is responsible for making. It is impossible for a president to give adequate attention to all the issues under his purview, which makes selecting and supporting trustworthy teams paramount.

Charlie and I received late word that we could each have one guest as we accompanied the First Family at the Cape that day. I pulled my youngest son, Mitch, out of school to join us, and his presence—and banter about how the Chicago Bulls were doing in the NBA playoffs—seemed to visibly relax the President. Although the Shuttle launch was postponed, we provided an astronaut tour, and watching Sasha and Malia (as well as Michelle and her mother) listen to Janet Kavandi describe her space missions reinforced for me how human spaceflight had the power to be inspirational.

In his memoir, Obama reflects on the impact meeting a female astronaut had on his daughters and expresses fond memories of the visit. The former President notes how he highlighted STEM education during his tenure, adding, "I'd also encouraged NASA to innovate and prepare for future missions to Mars, in part by collaborating with commercial ventures on low orbit space travel." That is his entire documented remembrance of NASA during his eight-year tenure.

President Obama's brief mention may portend progress from the troubles that inspired Vice President Quayle to dedicate an entire chapter of his own memoir to his NASA travails, but I can't help but be a bit disappointed by our showing. I'm proud of the accomplishments we made, but I believe President Obama's decision to shift NASA to a more advanced, sustainable agenda could have made even greater strides sooner. I'd have placed a bigger bet on our hand, but I still believe history will remember the 44th President's positive impact on NASA as transformative.

7.

DARK MATTER

WORKING AT NASA ALLOWED ME TO LEARN MORE ABOUT ENGINEERING and science than I would have ever thought possible for someone with political science and economics training. While I discovered that brilliant people could explain just about anything, the most challenging of all the topics for me to grasp was astrophysics. Structural engineering, propulsion, planetary and Earth science, biology, chemistry, even astrobiology (okay, we made that up, but now it is a thing), seemed simple by comparison.

The Alpha Magnetic Spectrometer (AMS) is a mission proposed to NASA in the 1990s by Dr. Sam Ting, a Nobel Prize–winning astrophysicist. The instrument was sponsored by the Department of Energy and was designed to be attached to the ISS but was scrapped after the Columbia accident. Dr. Ting came to see me during the Obama transition team period to lobby for it to be put back on the manifest and to explain how the instrument detects antimatter in cosmic rays.

Dr. Ting is intense and the briefing was dense. It was hard to know which of my questions were silly or pertinent. When I asked him what dark matter was, Dr. Ting's explanation set me at ease. He told me that scientists could spend hours trying to explain it to me, but that dark matter is essentially everything we don't understand. We ultimately secured the funding, and I was thrilled to see the AMS become operational.

I do not know if Dr. Ting's description is scientifically accepted, but in my world, it made sense. After the experience I referred to the illicit and unethical behavior of a few senior leaders at NASA—actions I could never understand—as dark matter.

To me, there is no worse crime for a government employee than to use the public's money for his or her own purposes. It goes against a political scientist's laws of nature. I found it far too common for Agency employees to act entitled when deciding how tax dollars should be spent. My first example is Mars, and my story starts on Earth.

Addressing the challenges from global warming was a high priority in the Obama administration, so the transition team had been asked to give it greater emphasis in our planning. The first organizational change I recommended to Charlie was to separate and elevate the Earth Science Division out of the Space Science Directorate. Dr. Holdren supported the move, and at first Charlie seemed open to the idea. A few weeks later, when I raised the issue with him again, he told me he'd asked the head of the Science Directorate, Dr. Ed Weiler, and he didn't want to do it. No dice.

We'd discovered during our transition team review that NASA's Science Directorate was one of the areas where reforms were desperately needed. Much of space and Earth science was still bogged down with billion-dollar missions that took decades to build and stifled innovation. Strictly adhering to the National Academy of Sciences decadal surveys left little room to take advantage of new technologies or opportunities, leading the peer-review process to be used as a cudgel. Ed had opposed every idea we proposed, proving that military service isn't required to be a cup boy.

Dr. Ed Weiler was appointed by Mike Griffin to head NASA's Science Directorate in 2008 after spending thirty years in various science leadership positions at the Agency. We'd worked together under Dan Goldin and gotten along fine, although I steered clear of him as much as possible, given his penchant for rude and lewd commentary. Ed was a product of the bureaucratic environment, having perfected the skill of avoiding blame for problems that occurred under his watch.

If NASA was going to transition into a twenty first–century agency of innovation, we couldn't ignore the $6 billion science budget. When Charlie wasn't willing to replace Ed, I searched for workarounds. One of my more successful strategies was the revival of the chief scientist and chief technologist positions, which allowed us to recruit more creative leaders to breathe life into the system.

NASA's astronomy and astrophysics programs had been Ed's responsibility in some fashion since the mid-1990s, when the James Webb Space Telescope was initiated. Designed to be the follow-on to Hubble, the program was estimated to cost $500 million and to be launched in 2007. During Ed's tenure, the program's cost increased twenty fold and was delayed more than a decade. Its continued existence was a consequence of spreading work across the country—especially in Maryland, home to Senator Barbara Mikulski. Webb's billions of dollars in cost overruns wreaked havoc with the science budget and left nowhere to cover the shortfall other than planetary science. Within planetary science, the Mars program was the only program with that much money to give.

The Mars Science Laboratory was the most ambitious lander ever proposed and was supposed to be a year away from launch when I arrived back at NASA in 2008. When JPL's Charles Elachi had expressed his concern about cutting corners to make the 2009 launch window to the transition team, we'd all agreed it would be worth the wait and the gold—two years and $600 million dollars added to its previous $1.9 billion-dollar budget. Having experienced previous Mars mission failures at NASA, I knew enough to give the scientists and engineers every chance to succeed.

Dr. Elachi thanked me personally in his speech after Curiosity's successful landing on Mars four years later.

Decisions have consequences, and the combination of Webb Telescope and Mars Science Lab's billions of dollars in overruns eliminated the budget for the next Mars lander in the queue—a cooperative mission with ESA called ExoMars. Legend had it that Ed Weiler had already agreed to the mission with his counterpart at ESA, but on this side of the pond, the administration, through OMB, had been crystal clear about our inability to proceed with the mission. Ed didn't want to hear it, and Charlie preferred not to say no.

Formal Administrator correspondence was managed tightly at NASA. The process included a routine deputy sign-off. I signed off on 99 percent of the correspondence that came to me for review. Depending on the topic, there were sometimes ten or more signatures before mine. If I raised an issue, the process had to begin again. Knowing the mission was considered canceled by the White House, I was surprised to find a letter to ESA committing NASA to partner on ExoMars in my inbox to review for Charlie's signature.

I talked with OMB and confirmed the letter went against administration policy and should not be sent to ESA. Ed had gotten the Administrator's buy-in by suggesting the letter was just a placeholder to let ESA know we were still looking for the money. NASA is part of the administration and we couldn't just ignore their direction. Ed didn't want to admit to ESA that he'd spoken out of turn and overcommitted his government, likely planning to string them along while he made an end run to get the money from Congress.

Ignoring cost overruns on missions in his own directorate led to the lack of funding and late decision on ExoMars, but Ed was a bureaucrat who had learned to blame such things on others and used me as his scapegoat. He told anyone who would listen that I was personally responsible for canceling US participation in the ExoMars program.

In JSC's oral history interview with Ed shortly after his retirement from NASA, Ed was asked a question about why our participation in the mission was canceled. His response: "The reason it was done—I'll be brutally honest—was our Deputy Administrator, who was connected to the commercial world, Lori Garver. Oh, she was the biggest disaster in NASA history for science. She had an inside track to OMB. They would go around Charlie routinely and undercut him. He was trying to get a letter agreeing with ESA to work together, and they got it cut. They got that letter stopped. A lousy letter, they got it stopped. Lori Garver was one of the main reasons I left. I couldn't deal with that person." He added, "For us to shoot them in the face—after all the hard work they did to get their member nations to agree to do something they'd never done before, to commit to a series of missions—to have this arbitrary and capricious decision made not to go forward with the Mars program, I was just totally embarrassed. I couldn't take it anymore. It's one of the main reasons I left NASA. I couldn't take it. I was only sixty-two, but I could not deal with incompetence. I can deal with people who are technically qualified to make technical decisions, but not people like Lori Garver and the people at OMB to make those kinds of decisions."

Astronaut Jim Lovell, Tom Hanks, and Lori Garver at NASA Headquarters - July 1995. *NASA*

Lori Garver, Mike Collins, Buzz Aldrin, and Neil Armstrong, appearing on *Meet the Press* with Tim Russert - July 1999. *NASA*

Greeting President Clinton with a gift of an Apollo Moon rock during an Oval Office visit with the Apollo 11 astronauts - July 1999. *NASA*

Lori Garver and NSYNC's Lance Bass announcing the completion of their medical certification and initiation of training for a Soyuz flight in Moscow, Russia - May 2002. *Getty Images.*

Charlie Bolden and Lori Garver testifying in the Senate at their confirmation hearing - July 2009. *NASA*

President Obama greeting Lori Garver after his address at the Kennedy Space Center - April 2010. *NASA*

(Above) Lori Garver touring SpaceX's Hawthorn, PA, facility with Elon Musk - September 2010. *NASA*. (Right) Former astronaut and Senator John Glenn with Lori Garver at the 50th Anniversary of his Mercury 7 flight, Columbus, Ohio, February, 2012. *Lauren Worley*

Sir Richard Branson, Governor Richardson, Lori Garver, and Buzz Aldrin arrive at the runway dedication ceremony of Spaceport America - October 2010. *NASA*

Lori Garver speaking with students at the NASA New York City Educational Forum - March 2011. *NASA*

Lori Garver introducing her son Mitchell to President Obama at the Kennedy Space Center - April 2011. *NASA*

Lori Garver and Chris Ferguson, commander of STS-135, standing under the Atlantis immediately following the final Space Shuttle flight - July 2011. *NASA*

Delivering the Space Shuttle Discovery to the National Air and Space Museum, with Jack Dailey, director of the museum; Wayne Clough, Secretary of the Smithsonian Institution; Ray LaHood, Secretary of Transportation, and other dignitaries - April 2012. *NASA*

Gene Cernan, the last man to walk on the Moon shares a thought with Lori Garver, as Senator Kay Bailey Hutchison (R-TX) and Lori's husband David Brandt look on, while attending the memorial service being held for Neil Armstrong, the first man to walk on the Moon, at the National Cathedral. September, 2012. *NASA*

Lori Garver speaks at the Los Angeles Airport, in front of the Space Shuttle Endeavor as it arrives to be delivered to the LA Science Museum - September 2012. *NASA*

Attending SpaceX's Falcon Heavy launch with NASA's Phil McAlister and Marc Timm at the Kennedy Space Center - February 2018. *David Brandt*

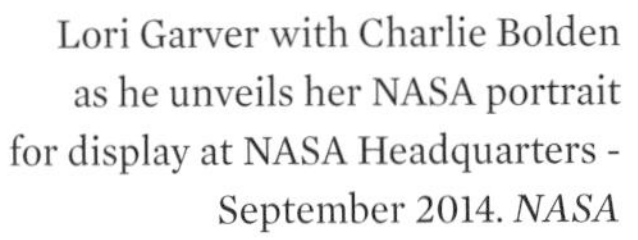

Lori Garver with Charlie Bolden as he unveils her NASA portrait for display at NASA Headquarters - September 2014. *NASA*

Lori Garver and Jeff Bezos at the National Air and Space Museum - September 2016. *David Brandt*

Ed's description of me as a Bond-like villain *connected to the commercial world* is a nonsensical attempt to deflect his own complicity. His accusation that I was going around Charlie was, in reality, what he had done himself. Ed was the person responsible for shooting ESA in the face. He'd run the risk of making commitments of taxpayer funds without authority, likely intending to trap the US government into doing what he wanted. Ed had been learning how to be a practitioner of the dark arts for thirty-five years, and had become an expert.

It is not only unethical for government employees to commit government funding without authority but also a criminal violation. The Antideficiency Act (known as the ADA) "prohibits federal employees from making or authorizing an expenditure from, or creating or authorizing an obligation under, any appropriation or fund in excess of the amount available in the appropriation or fund unless authorized by law." Making commitments on behalf of the US government without having the authority to do so is a crime.

Another instance of this unauthorized funding game nearly happening at NASA had everything to do with the intensity of the battle against terminating the Constellation program.

It is a longstanding and universal requirement in large government procurements to ensure that enough funding has been budgeted to cover any losses that the contractor might incur in the event of a program cancellation. The amount of such funding is called termination liability. Given the annual nature of government appropriations, this requirement is the method by which the government ensures that it has not exceeded the amount of budget authority provided by Congress. It also helps to undercut the natural tendency to pour good money after bad by saying, "It would cost more money to cancel it now."

When the Obama administration recommended canceling Constellation, the contracts were reviewed to determine whether the proper amounts of funding had been withheld. NASA CFO Beth Robinson and her team discovered that most of the prime Constellation contracts didn't have the usual language providing termination liability. Without legal protection, the contractors were responsible for covering their own close-out costs, but had not withheld any funds. A conservative estimate of their liability for termination costs at the end of fiscal year 2010 totaled close to $1 billion.

By allowing companies to spend all of their money on hand (and therefore not reserving any for termination liability), the Constellation program appeared less expensive—which was obviously the goal of the previous administration—but levied unreported risk on each of the three prime contractors involved. This strategy was entirely consistent with what the transition team had found: the goal was to rush contracts and do everything possible before a new administration would realize what was happening. When NASA notified Congress and reminded the companies of their liability, the shit storm over canceling Constellation intensified. Proceeding with termination according to the contract language was likely to lead to shareholder revolt, SEC investigations, board actions, and management shake-ups.

The companies had taken on hundreds of millions of dollars of risk, somehow believing the government would bail them out, but obviously NASA had no such budget or contractual obligation to do so. Industry representatives argued that even though the requirement is levied on all government contracts, they had been told by NASA that they would not be responsible for covering such costs.

The contractors repeatedly claimed that NASA representatives had assured them orally in meetings that termination costs would be covered by the government at no cost to them. An investigation to determine why the industry teams were under this impression eventually found an email between lower-level NASA officials and industry contracting officials with an oblique reference to allowing for termination costs.

I was incredulous that any non-contractual assurance would have been enough for company CEOs to take on hundreds of millions of dollars in liability on behalf of their shareholders—even if the promise had been made by the NASA Administrator himself. It certainly wasn't credible that they would have taken such a risk based on a low-level contracting official's email. I saw it as yet another reason Mike Griffin fought so hard to keep the transition team from "looking under the hood."

Not everyone supported Beth's recommendation to escalate the issue, but failing to report an ADA violation is a felony. A federal employee who "knowingly and willfully" violates the act "shall be fined not more than $5,000, imprisoned for not more than two years or both." We reported the potential violation, but the Administrator and Associate Administrator determined that there was no reason to search for additional documentation or look into the matter of who in NASA's senior ranks had made oral representations.

Members of Congress—those constitutionally charged to steward taxpayer dollars—defended the contractors. Raising the issue escalated the battle and led to the congressional push to adopt authorization terminology that kept the contracts in place. Congress did the industry's bidding, doing everything possible to force NASA's hand in the matter. Extending the Constellation contracts had the designed impact: it made the issue of termination liability moot. Several NASA and Hill staffers who worked to assure the contracts remained in place later moved into senior positions at the aerospace companies they defended.

Ultimately, the contracts were renegotiated to give the companies not only additional funds for termination costs but many billions more. Since the contracts weren't actually terminated, the Government Accountability Office agreed not to file charges but put out a report the next year with the title, "NASA Needs to Better Assess Contract Termination Liability Risks and Ensure Consistency in Its Practices."

Government service requires integrity and many of the behaviors I saw, in my view, should not have been tolerated. Others had a different level of tolerance, which perpetuated the behavior. Even when reported through established channels, the issues were rarely acknowledged or corrected. Escaping the trappings of power is sometimes harder than escaping gravity.

• • •

Another egregious instance of dark matter that required my intervention started with a request for NASA to transfer $90 million to the National Reconnaissance Office (NRO) in late 2010. The NRO is the US agency responsible for designing, building, launching, and operating spy satellites for the intelligence community. The funding transfer was to cover a share of an NRO facility being built at Cape Canaveral, but OMB couldn't determine what use NASA had for the resource. NASA had an excess of infrastructure at the Cape, and knowing I had the proper clearances, our budget examiners asked me to look into the matter.

I had intentionally kept from inserting myself into NASA's affairs with the military or intelligence communities. We had an abundance of former military leaders with greater interest and experience, including the Administrator and Associate Administrator. I maintained the clearances my position required, but only when I was invited did I attend meetings in the room reserved for classified information review and discussion—the Secret Compartmentalized Information Facility, known as the SCIF. Charlie made it a practice to visit the SCIF on a weekly basis to read intelligence reports, and I'd asked him to let me know if he ever came across anything I should make a point to read. He never raised any issues with me, so either there wasn't anything meaningful in the reports or there was nothing he thought I needed to know.

NASA is a civilian agency by law, and as such, has a mandate of transparency. The Obama administration was especially committed to transparency and restoring public trust in government. Very little NASA does is, or should be, classified. To me, the SCIF represented a holdover of the patriarchal, military culture at NASA, and I jokingly referred to it with my staff as the tree house—reminiscent of the forts boys built and filled with cigarettes and girlie magazines.

Avoiding the SCIF was impossible, since there was no way to know whether or not you needed information without going there to learn what you might be missing. A fundamental requirement for holding meetings in any SCIF is that actual classified information is being discussed. The requirement was often unmet—another way bureaucrats chose to control information. I raised my concern at the conclusion of a meeting where we hadn't discussed anything classified, and the reaction was that the meeting's location was necessary because I could have asked a question that would have required a classified response. My raising the issue made it even less likely I'd be invited to more briefings in the SCIF—which was fine by me.

The NRO question—what use did NASA have for a new facility at the Cape?—came to me because there was a lack of trust between the OMB staff and many other NASA leaders. Being trusted by OMB is considered by some as unseemly, but the opposite is true. The administration is a team, and each team member has an important, unique role. Of NASA's nearly 20,000 employees, about twenty are actually appointed by the President and therefore called "political," but civil servants are also part of the administration. Being a trusted team member of the administration assures that NASA has a seat at the table. Trying to keep information from OMB is like a company trying to outsmart its board of directors. You do so at your peril.

I was unable to uncover any planned use for a new $90 million high bay in Florida, and I relayed that finding to OMB, who denied the funding transfer.

A few days later, I received an invitation to visit with NRO Director Bruce Carlson at his offices in Chantilly, Virginia. Charlie and Chris accompanied me to Chantilly but awkwardly deferred to me in the discussion. Several other NASA staff were in the conference room when we arrived, and as I began to explain that we'd been unable to determine NASA's planned use for the facility, a sympathetic member of the NASA delegation caught my attention and mouthed the word "Altair."

Altair was the name of the notional lunar lander for the Constellation program that wasn't in the budget, even before the program was canceled. I apologized to Bruce for NASA's miscommunication but confirmed NASA had no use for the facility. I never learned who promised $90 million of the budget without approval, but whoever it was had likely violated the Antideficiency Act.

Not wanting to excuse ourselves after just five minutes, or to waste the opportunity, I raised another topic of mutual interest—shared launch vehicles. Key NASA and Senate leaders commonly used launching NRO satellites as a justification to build the SLS, so I wanted to run the question to ground. I'd raised the same issue of SLS use in NASA's quarterly meetings at the Pentagon. The Air Force, Space Command, and Strategic

Command had universally and defiantly said "no, thank you"—without the thank you. Charlie, Chris, and others conveniently ignored these discussions and continued to include launch of military and intelligence satellites in their talking points to justify SLS.

When I asked NRO if they had any interest in using the vehicle, their response was immediate and unanimous: no. NRO deputy Betty Sapp offered a reason for their quick response—their satellites had precision instruments that could not withstand the dynamic environment of launching on large solid rocket motors. There it was: the very element of the rocket being forced on NASA by congressional leaders, people doing the bidding for self-interested contractors, had limited the types of payloads that could be launched on the vehicle. Chris Scolese worked closely with the NRO (and subsequently became the NRO director in 2019), so he must have known their constraint, but he again chose to look the other way.

A handful of senators had raised national security as a reason the Constellation program couldn't be canceled in 2010. Senate Budget Committee Chairman Kent Conrad (D-ND) claimed, "There are classified discussions that we can't go into here, with respect to this initiative, but I'd say to my colleagues this is absolutely essential for the national security that this go forward." Senator Nelson made similar claims focused on the Ares I rocket and SLS, saying its development was critical to maintaining the US capacity to build the large solid rocket motors used on strategic missiles and satellite launchers.

The real issue here was the availability of solid rocket motor propellant for military use. The solid rocket motors used on the Shuttle and planned to be modified for Constellation and SLS needed much larger quantities of propellant than the military's projectiles. Senators Nelson, Conrad, and the Utah delegation insinuated that if NASA ceased to be the primary customer for the propellent, US production would also cease, leaving our nation's ICBMs and Minuteman rockets unable to be refurbished.

I thought that claims of national security threats should be taken seriously and chartered a study with Department of Defense on the issue. A yearlong interagency review revealed that NASA's investment in solid rocket propellent had conveniently reduced the cost of the military's purchase by $30 million. It wasn't a national security issue; it was a budget issue. The information was moot by the time the study was complete, since NASA had already agreed to continue to extend the contracts for solid rocket motors on the SLS. The experience was yet another reminder of the lengths the opposition was willing to go in pursuit of their agenda.

Charlie and I have been referred to as a Team of Rivals. While the comparison is somewhat analogous, unlike President Lincoln's appointments of Seward, Chase, and Bates, President Obama hadn't appointed Charlie or me to offer competing views to his own or to each other. In response to a question in a 2016 interview about whether he thought Team of Rivals was a fair way to characterize our relationship, Charlie said, "No, we didn't function as a team; our leadership was dysfunctional. There were people loyal to her, and others to me."

I was disappointed by Charlie's characterization. I thought the political team was loyal to the President, the administration, and NASA—including the Administrator. On the occasions where Charlie's views were opposed to the administration's, we all had to make hard choices, but on most matters I thought the team's loyalties were aligned. I considered Charlie to be our team leader in all situations unless he was doing something in opposition to the President or his policies. If there was a disconnect between them, I believed my allegiance was ultimately to the President.

For most Deputies, this doesn't become an issue because the Agency head aligns with the White House on significant policy issues. I viewed Charlie's opposition to administration policies as peculiar. A comparable situation might be if the head of the National Institutes of Health (NIH) didn't support the president's policy to conduct fetal tissue research. Say the NIH director decided not to request money for the research in their budget, retained researchers who opposed the policy, and asked them to back-channel information to the Hill to undermine the practice. Whether it was Charlie's intent or not, this disconnect became a significant challenge to manage during NASA's 2011 budget process, and unfortunately it was not an isolated incident.

The best solution would have been for Charlie and the administration to align themselves more effectively, but Charlie's initial trusted advisors—the cup boys—did their best to keep that from happening. When I found myself in the middle, I tried to delicately raise the issue through the channels at my disposal and was generally told to try harder to get Charlie onboard.

The administration gave NASA a light touch compared to Cabinet Agencies whose leadership was often tasked with more political assignments. The previous White House had even told my predecessor how to dress and wear her hair, something I could never have imagined in either the Clinton or Obama administrations. The lion's share of the pressure on Charlie wasn't coming from the White House. Chris Scolese, Mike Coats, and others lobbied Charlie to take positions that opposed the administration for their own advantage.

The most bizarre incident of disagreement between Charlie and the President came in the summer of 2010 when, during a visit to Qatar, he was interviewed by Al Jazeera and said that the top three goals he was tasked with by President Obama were to help re-inspire children to want to get into science and math, to expand NASA's international relationships, and, "perhaps foremost, to reach out to the Muslim world... to help them feel good about their historic contribution to science... and math and engineering." The video of him giving this "feel good" statement is stunning and went viral over the next few days. When a reporter asked about it during a White House press conference, Press Secretary Robert Gibbs responded that "Charlie Bolden was wrong. The President never said that."

The morning after Robert Gibbs's assertion, I found Charlie in his office fuming. He wanted Gibbs to retract the statement because "he wasn't going to lie for them." I expressed my concern about the situation and asked if he could remember when the President said this to him, walking him through the handful of times they'd met without me. Considering it further, he realized the President hadn't said it and decided it must have been Secretary of State Clinton. I was in the only meeting he'd had with the Secretary at that time and said I hadn't heard it. He then remembered that it was someone else he'd met with from the State Department, but he couldn't remember his name.

Charlie's recollection that it hadn't been the President, or Secretary of State—or anyone in any formal capacity—that had tasked him with Muslim outreach goals, hasn't been made public to my knowledge. I do not believe it was Charlie's intent, but former President Obama is still ridiculed for supposedly telling the head of NASA that his number one goal for the space agency was to inspire the Muslim world.

Early in our tenure, after hearing Charlie make negative comments about the administration in a larger setting, I privately suggested he modify his comments, given we were part of the executive branch. He didn't agree and said we worked for the Congress and the administration equally, since we were appointed by the President and confirmed by the Senate. Charlie's confusion over this distinction was surprising. We had what I viewed as a positive discussion over where NASA fit into the government. At one point his questions led me to pull a piece of paper out of my binder and draw a triangle to explain the three branches of government. At the end of the discussion Charlie appeared grateful for the information but continued to side with the cup boys more than the President.

Charlie's reluctance to make staff changes or accept new political appointees made it challenging to build a leadership team that worked toward a common purpose. I did my best to overcome his reticence by making structural changes and creating new positions. These maneuvers did more damage to my relationship with Charlie, but the people they allowed us to recruit became critical to our success and included Laurie Leshin, Bobby Braun, Mike French, and David Weaver. It became easier to drive meaningful change when more openminded NASA leaders started populating senior positions. The team I worked to cultivate eventually became Charlie's most trusted advisors.

New members of the leadership team started embracing fresh ideas, and their emphasis on innovative practices allowed NASA to make progress in several areas. Greater use of partnerships, data buys, hosted payloads, reusable suborbital science missions, green aviation, prizes, and subleases of government infrastructure were concepts that evolved over time with varying degrees of success, taking hold when other NASA leaders saw their value.

The Kennedy Space Center eventually became so overzealous in their effort to shed the costs of existing facilities, they were ready to sign a sole-source agreement with SpaceX for Launchpad 39A. When I heard about the deal, I pushed KSC to hold a competition instead of just transferring the pad to SpaceX.

Other efforts followed the pattern of Sisyphus. I'd work for months or even years on a project, only to have it overturned—sliding back down to the bottom of the hill.

Stimulating the development of commercial reusable suborbital research vehicles was a boulder I was determined to push. Working with like-minded individuals at the Agency, NASA began a program in 2009 to support the effort with $2.5 million. In 2010 we established a five-year program with an annual budget of $15 million. Funds were to support experiments and researchers who could utilize the emerging capability, but by 2012, of the twenty-one experiments selected, fourteen were awarded to fly on existing platforms—NASA balloons and airplanes. The program offices were happily spending the new money, as long as they could keep doing what they had been doing.

Most researchers needed to be onboard the vehicles to conduct their experiments, but NASA didn't want to take that risk, restricting its awards to autonomous research. I eventually convinced Charlie to remove the restriction in 2013, and I made the announcement at a conference that June. Within weeks, the cup boys got Charlie to reverse his decision. The next NASA Administrator eventually put the policy in place, but having it on the books five years earlier would have expanded the addressable market for the fledgling industry when they were building their business cases.

Unexpected dark matter emerged around a research project I championed. OMEGA—Offshore Membrane Enclosures for Growing Algae—was a project focused on green-biofuel development for aviation. The team created an innovative method to use algae to clean wastewater and capture carbon dioxide to produce biofuel without competing with agriculture for water, fertilizer, or land. It literally turned shit into fuel and fresh water. I was impressed by a demonstration presented by our aeronautics staff at the Ames Research Center in California and wanted to support their request for an additional $5 million in funding to meet their next milestone. I discussed OMEGA with other potential government collaborators, including the undersecretary of the navy, who was extremely positive about the concept and interested in our results.

Before expressing my support for the funding extension to Dr. Jaiwon Chin, NASA's Director of Aeronautics, I ran it by Charlie, who gave me his blessing. Jaiwon agreed to continue funding the alternative fuels project, but the OMEGA team later learned they wouldn't receive the extension. When I circled back with Jaiwon, he said he'd talked to Charlie, who told him he didn't need to provide the funds. I went to the Administrator to ask him directly, and he admitted to changing his mind after talking to a former colleague at Marathon Oil who didn't see a lot of promise in the alternative fuel project. Charlie's previous board service and continued ownership of a half a million dollars of the company's stock meant he had a formal conflict of interest restriction on activities related to Marathon Oil. His contacting them about OMEGA was not allowed by his formal restriction. The general counsel, who was a long-time marine friend of Charlie's, didn't consider the infraction to be serious, but the Inspector General took a different view.

The IG found Charlie in violation of his ethics pledge and directed he recuse himself from any decisions related to the project. The IG's report was made public, and Charlie received additional ethics training. Believing the issue was settled, I prepared to approve the $5 million project. But Charlie directed a new process for OMEGA that went around the existing chain of command. He gave responsibility over the project to the Associate Administrator instead of the Deputy. Making this decision likely violated his recusal by reinserting himself in the issue. And guess what? It kept the project from moving forward. NASA's investigation of an alternative way to produce aviation fuels that could reduce the release of greenhouse gases, decrease ocean acidification, provide fresh water, and enhance national security, was a boulder I eventually had to leave at the bottom of the hill.

The tendency of a few former senior military officials and astronauts to behave as if the rules didn't apply to them ran counter to my own sensibilities. There seemed to be a sense that government service entitled them to benefits well beyond what I felt was appropriate. I raised concerns over a few practices I considered to be inappropriate uses of government funds, such as expanding all astronauts' lifetime healthcare benefits to their extended families, use of government airplanes for what I viewed as non-essential excursions, and government funding of the NASA leadership team's spouse travel to attend the Administrator's holiday party (this was ultimately not approved by the general counsel). Questioning such perks was viewed by others in NASA's leadership as impertinent, and the greater offense. The perpetual cozy system that shaped the behavior largely remains in place.

Much of the dark matter I saw seemed to be learned behavior stemming from expectations set over time for exalted leaders in government. A 2020 IG investigation found that Charlie had used the services of his former administrative assistant to manage his private consulting activities and coordinate additional services from other NASA personnel for almost two years after leaving the Agency. The IG report notes that when he was interviewed, Bolden initially denied receiving such support but after being shown evidence including his many email requests, acknowledged what he'd done and that it was his "error in judgment." Charlie told the investigators that "one of my biggest disappointments after leaving NASA was how little support I received." Having been an astronaut and marine general, his expectations seemed to include government funding of professional speech writing and administrative support for his post-employment business activities. It turned out to be true. The IG decided Charlie didn't have to reimburse the government for the services that were inappropriately provided, based on his long service to the nation.

• • •

I didn't have regular audiences with President Obama, but I remember every word he ever said to me. When he and his family were at the Cape for the scrubbed Shuttle launch, he pulled me aside to speak privately. He said his people had told him that I'd been the one taking arrows for the team, and he wanted me to know how much he appreciated it. That he knew I was trying to push valuable boulders up the hill meant everything to me.

Later, right before the 2012 election, I was in the Oval Office for a photo-op with other members of the Deputies Council. The President asked me to stay behind for a moment and offered me an apology for not having gotten all the big things done that we wanted to at NASA. His face looked drawn, and I was mortified that the Agency had been a burden to him politically. He said, "We'll just have to get it all done in the second term." He told me to keep my chin up. I thanked him and said he should do the same.

We inherited the human spaceflight program at a time when Space Shuttle retirement was preordained, and its planned replacement program was irrecoverably off track. No matter how it was handled, it was going to be a huge challenge, even in the best of circumstances. Having the unwavering, aligned support of the administration at every level was necessary to get human spaceflight on a sustainable path, and the early progress we made was only possible because we initially had that from the Obama administration.

David Weaver, NASA's extremely talented head of communications, described the public affairs challenges of retiring the Space Shuttle without an immediate replacement as surfing on a tsunami. He recognized that the tidal wave of public and institutional support for the Shuttle was too challenging for our nuanced message about the value of our proposal for human spaceflight to convey at the time. He said our goal should be simply to keep our head above water. If we did that, when the tsunami was over, we could pick ourselves up and save human spaceflight—and possibly NASA—from drowning under the weight of the past. I trusted David's expertise, but even I could not have imagined the intensity of the comeback.

8.

RISE OF THE ROCKETEERS

TRANSFORMATIONAL SHIFTS IN GOVERNMENT SPACE SYSTEMS REQUIRE A coordinated, aligned force—the right combination of people to advance policy, technology, and investment. Disrupting a paradigm as ingrained as the space-industrial complex means risking one's career and financial future to drive change. The tremendous challenges and opportunities related to utilizing outer space have attracted many of the best and brightest minds, and the world is a better place because of their efforts.

SpaceX, Virgin Galactic, and Blue Origin are the most visible private companies advancing more accessible human spaceflight today. Each was backed by significant funding from their founders before earning government or commercial contracts. Elon Musk, Sir Richard Branson, and Jeff Bezos have risked their reputations and funneled a portion of their own fortunes into these companies in order to realize their dreams in human spaceflight. They are not the first to do so.

The potential for communications satellites to deliver value beyond NASA and the government was recognized early, which led to their privatization through COMSAT and INTELSAT in the 1960s. These initially quasi-government organizations contributed to the evolution of a burgeoning, profitable telecommunications market that transformed society through instantaneous communications. Private entities incentivized by national governments and international organizations helped create the infrastructure that first allowed us to transmit signals, then voices, pictures, and video, and eventually the internet to anyone else on the planet anywhere, in real time.

The promise of new satellite constellations in the 1980s and '90s that fueled traditional aerospace industry interest in NASA's Reusable Launch Vehicle program also inspired the formation of smaller private rocket companies focused on capturing this expanding market. Many of these ventures had long-term visions of carrying people to space. Some early companies bore more fruit than others, but all helped to lay the foundation for the successes that followed.

A perceived market for space tourism has been talked about and pursued for decades. Five years before Apollo 11 touched down on the Moon, Pan Am Airlines began its "First Moon Flights Club," which garnered nearly 100,000 applicants before shutting down its waiting list in 1971. In 1985, an exotic travel company called Society Expeditions announced it would begin flying people to space for $52,200. The firm took in public deposits of $5,000 per seat, but the spacecraft was never completed, and Society Expeditions eventually returned the money.

The concept of human spaceflight wasn't confined to tourism. Economists, conservationists, and futurists began considering the value of moving mining and heavy industry, such as energy production, off Earth in the 1970s. Studies examined how free-floating structures could support large populations and activities that would benefit from microgravity and assure the survival of humanity on Earth and beyond. This early space ideology advanced transformational thinking related to self-sustaining space habitats. Gerard O'Neill's ideas not only inspired the founding of the L5 Society in the '70s but also attracted a cult-like following that grew to include Timothy Leary, known for researching the effects of recreational LSD, and Princeton student Jeff Bezos.

One of Timothy Leary's friends, George Koopman, started American Rocket Company (AMROC) in 1985. Like others in the field, the company was founded to reduce the cost of space transportation and capture the satellite launch market. Koopman was a celebrity space pirate by the time I met him in the mid-1980s, and his Timothy Leary and Hollywood connections added glamor to the early space movement. George tragically died in a car crash at the age of forty-four, just four months before AMROC's first launch attempt. The company eventually filed for bankruptcy.

AMROC's leaders and intellectual property moved on to other similarly motivated efforts that have contributed to some of today's leading commercial space companies. The lineage of serial rocket designers and entrepreneurs is, indeed, close-knit and complex. A handful of space pirates developed a dozen early companies focused on achieving reusable, cheap access to space; companies like Rotary Rocket Company, XCOR Aerospace, Kistler Aerospace, SpaceDev and Space Services Inc. Each of these ventures in turn grew more designers and investors.

It has been a privilege to have known and worked with many of the individuals who invested their own money to advance sustainable space transportation in the 1980s and '90s. Most didn't go into the business just to make money, and some lost their shirts in the process. The not very funny joke at the time was, "How do you become a millionaire in the space business? Start as a billionaire."

The perceived growth of a massive satellite market that drove much of the early private investment in launch vehicles in the 1990s, evaporated due to a number of unforeseen events, most notably the burst of the dot com bubble. Still, the early projects laid a foundation for later successes in areas such as commercial policy planning, new technologies, advances in satellite development, experience gained by individuals who remained in the industry, and even facilities built to support the activities. The unexpected ingredient was the entrance of billionaires who, thankfully, hadn't taken the joke seriously.

The first billionaire rocket developer I ever met was Andy Beal. Beal founded his self-named rocket company in 1997, having made his fortune in Dallas-based banking and real estate. Like the others, Andy wanted to capture the much-anticipated market to launch communications satellites. Beal Aerospace attracted a mix of space industry talent from traditional aerospace, and I visited the facility in the late '90s and early 2000s as part of my NASA commercial policy review for Dan Goldin.

Andy Beal said he didn't need incentives from the government; he just wanted to be assured it wouldn't compete with him. After spending $250 million of his own money, just months after what were reported as successful engine tests, he shuttered his doors in 2001, blaming NASA's intention to unfairly compete with him by building its own rocket. The engine facility that Beal Aerospace built in McGregor, Texas, was later acquired by SpaceX, becoming the core of the company's propulsion testing.

A few hundred miles to the northwest, in the desert at the edge of Las Vegas, another billionaire had his eye trained toward space. Like Andy Beal, Robert Bigelow made his fortune in real estate. His desire to build wealth, he said, was fueled by his interest in space—as well as his long-standing belief in extraterrestrial beings. He founded Bigelow Aerospace in 1999 with the mission to "provide safe and low-cost commercial space platforms for low Earth orbit, the Moon and beyond." Bigelow acquired the licensing rights to a NASA technology for expandable space habitats as the basis for his space platforms.

Rather than launching a space station piece by piece, as NASA and its international partners were doing, Bigelow Aerospace's expandable modules could launch all at once and then expand once in orbit, an undeniable advantage. I visited the facility in 2011 and believed their space habitats might be suitable to use as additions to the ISS. It took several years, but NASA eventually awarded the company an $18 million fixed-price contract to build BEAM—the Bigelow Expanding Activity Module—which launched to the ISS in 2016. The module is still being used for pressurized storage at a cost more than an order of magnitude cheaper than any other Station module.

Robert spent more than $350 million out-of-pocket to fund the company, which had 150 employees at its peak. In spring 2020, as the COVID-19 pandemic began to surge across the United States, the company's employees were laid off and Bigelow Aerospace remained shuttered.

Even farther west, a more well-known billionaire took an interest in space. Paul Allen famously became a multibillionaire at just north of thirty years old, having made his fortune after founding Microsoft in 1975 alongside Bill Gates. Two decades later, Paul turned his focus to how technological innovation could be extended beyond the atmosphere. Paul cofounded Vulcan with his sister Jody, and their initial space venture evolved into the first privately developed reusable spaceship, known as SpaceShipOne. Paul provided $25 million in funding to win the $10 million Ansari X-Prize, and he selected legendary experimental aircraft builder Burt Rutan as his partner and lead developer.

Flying out of its base of operations in California's Mojave Desert, SpaceShipOne conducted back-to-back flights to suborbital space in September and October of 2004—clinching the X-Prize. It took nearly twice as long as anticipated, but to many of us, it was a tremendous milestone. My husband and I made the trip out to the desert to see the prize-winning flight in person and were rewarded by watching what we all assumed was history in the making. Along with hundreds of others, we saw Sir Richard Branson announce he'd formed a joint venture with the team after the winning flight.

Paul invested in even larger space launch projects over the years, but none became successful operations before his death in 2018.

The entrance of Sir Richard Branson to the commercial space billionaire club was especially welcome news. Branson brought not just his wealth, but an unrivaled public brand and glamor previously missing from the industry. Branson named the new venture Virgin Galactic, establishing a business plan to fly "tourists" regularly to and from the edge of space on suborbital flights. The company signed up nearly a thousand people at a ticket price of $250,000, to spend a few minutes weightless and view the curvature of Earth. Virgin Galactic partnered with the state of New Mexico to build Spaceport America to serve as the hub of their business. I represented NASA at the ribbon cutting ceremony for the Spaceport's runway in 2010, and along with the hundreds of other well-wishers in attendance, believed success was finally right around the corner.

Richard is the most naturally charismatic of the billionaire space barons in my view. He exudes enthusiasm for both space and humanity. Our conversations have been memorable and meaningful. We share many progressive views unrelated to space and I've always appreciated his willingness to speak out on important social topics. Being a senior political appointee means carrying the baggage, good or bad, of the President. In the aerospace world, being tied to Obama didn't typically bring many friends in corporate CEO ranks. Richard follows US politics closely and we bonded over our admiration for President Obama. Our private conversations have been as one might expect, raucous and delightful. I joined Richard and his son (and George Whitesides) for dinner one evening in New Mexico after visiting the Spaceport, and found Sam Branson to be similarly engaging. Spending time with people around their children offers a glimpse into their more private nature, and the Branson family exudes a close bond.

Flying people even to just the edge of space has proven more difficult than anticipated. Virgin Galactic's spacecraft development suffered a pair of deadly accidents which significantly delayed their progress. In July of 2007, three workers were killed in an explosion at a test site in Mojave, California, during the loading of propellants. Since it was a flow test—not planned for ignition—it had mistakenly been viewed as inert, so all the employees hadn't moved behind the traditional safety barrier. Those who watched the test from behind the barrier were not injured. The accident set the program on its heels and caused a complete review of safety processes delaying future testing.

In December of 2014, two pilots strapped into SpaceShipTwo on its second powered test flight. After the vehicle's planned drop from its carrier aircraft—White Knight Two—the rocket's engines lit to power the spacecraft, but it broke apart eleven seconds later. One of the pilots was thrown free and survived his parachute landing, but the copilot perished. The investigation pointed to the copilot prematurely releasing the feathering mechanism needed to stabilize the spacecraft later in the flight plan. Richard, George (then CEO), and the entire Virgin Galactic team were devastated by the loss.

Fully recovered as of this writing, Branson and Virgin Galactic have raised several billion dollars since its founding in 2004, and they took the company public in 2019. It looked like Virgin Galactic had the corner on the suborbital tourist market for the first few years, but an even more well-heeled competitor had other plans.

• • •

The Moon landings inspired millions, but few can say they turned their space-bound dreams into their own rocket company like Jeff Bezos. Jeff distinctly remembers the Apollo 11 landing, even though he was just five years old. While his space company does not yet have the global recognition his primary source of wealth does, time and again, Jeff has called space his most important work.

Jeff, as a newly minted billionaire at the turn of the century, founded Blue Origin just as his book-selling business Amazon began its surge into one of America's most valuable and wide-reaching enterprises. While Jeff, Elon, and Sir Richard may have founded their companies around the same time, their focus and work cultures are distinct. Blue Origin's motto—Gradatim Ferociter—is Latin for "step by step, ferociously." Blue likes to reference the "Tortoise and the Hare" fable in their competition, hoping that slow and steady wins the space race.

Blue Origin remained small and largely quiet for its first decade, especially by comparison to SpaceX and Virgin Galactic. The company focused on testing rocket technologies in its early years, beginning with low-altitude, jet-powered test vehicles first in Washington state and later in Texas. Like Virgin Galactic, Blue Origin's commercial rocket system only reaches the edge of Earth's atmosphere today, but Jeff sees the suborbital rocket system as the first small step in a grand vision of moving Earth's manufacturing and other industries off the planet.

Twenty years after founding Blue Origin, the company has grown to have more than 3,500 employees, with numerous development projects underway. When the company began test flights of its first booster, called the New Shepard, they aimed to launch passengers in 2015. As others in the space industry have discovered, success often takes longer than expected.

Continuing to name their rockets after famous former astronauts, Blue Origin announced New Glenn in 2016. The enormous rocket was designed to dwarf even SpaceX's Falcon Heavy rocket. Its initial flight was planned for 2020 but has since been delayed to at least 2022.

Jeff unveiled the Blue Moon lunar lander at a highly publicized media event in 2019. He declared the Moon to be a "gift" given to humanity, envisioning it as a hub for in-space manufacturing due to the relatively low energy required to launch materials from the surface. The company partnered with a collection of large aerospace contractors to bid for a multibillion-dollar NASA contract to land astronauts on the Moon. It seemed like a brilliant move at the time.

Jeff's ultimate vision, as inspired by Gerard O'Neill, is "having millions of people living and working in space to benefit Earth—and moving industries that stress Earth into space." Blue Origin revealed in 2020 that it was looking into building a line of orbital habitats and, in 2021, announced its partnership with another innovative space company, Sierra Space (formerly Sierra Nevada). The team says the habitats are fundamentally different from the ISS, describing its space stations as destinations for people to visit and laboratories to conduct science.

I first met Jeff Bezos when I returned to NASA in 2009. Blue Origin was keeping a low profile in those days, but I was thrilled when they reached out to schedule an introduction. Jeff flew to DC to sit down with the NASA Administrator and me to tell us about his company and his plans for the future. Jeff wasn't the richest person in the world yet—he ranked 18th—but he was significantly higher on the list than Elon Musk, and his wealth made NASA take him seriously.

My discussions with Jeff have always felt like talking to a friend I've known for years. He's relaxed, inquisitive, and hilarious. During our initial meeting at NASA, Jeff was open about his plans for Blue Origin and made an offer for us to visit his manufacturing facility in Seattle. Charlie didn't express interest, but I accepted his invitation immediately.

Touring Blue Origin with Jeff was impressive. Not only was the factory and scale of the operation remarkable, his knowledge of each function and employee was extraordinary. Jeff loves to tell stories that make an important point simply. My favorite from my first tour was about engine-cleaning materials. Cleaning residual fuel from engines after test flights was a hazardous, expensive activity requiring toxic agents in clean rooms with extensive precautions. Focusing on reusability required reducing the cost and streamlining the activity, which led to an unexpected and innovative solution to the task of cleaning rocket engines—lemon juice. Jeff delights in telling the story of how he became the world's largest customer for the citrus extract.

Touring SpaceX with Elon was similar in many ways. Both men convey a deep knowledge of the operations, and employees don't overreact to their presence. You get the sense that seeing them walk around is a common occurrence. The difference between the tours and those at traditional aerospace contractors is extreme. The activity level at SpaceX and Blue is fast-paced. A spacecraft hanging in a high-bay under development regularly has six to eight people working on it at a time. Some people hang from scaffolding, some are on ladders—everyone has their own tool belt. Traditional contractor facilities were more often cavernous and quiet. A spacecraft in development typically had one person working on the vehicle, another standing nearby handing them their tools, and another watching with a clipboard.

I visited the Lockheed Martin facilities where the Orion capsule was being built several times and never saw anyone actually working on the spacecraft. Their message on my tours focused on how many different states had participated in providing parts or testing. During one visit to their Denver facility, the spacecraft had just returned from Ohio, where it had undergone a test they were now repeating. When I asked why they were doing the same test again, they said it was to assure nothing had loosened up in shipping. This made sense, and led me to ask why we had shipped it to Ohio in the first place. The senior executive leading my tour elbowed me and winked, saying they were doing their part to get the Ohio delegation onboard. As politely as possible, I suggested they should focus on building the spacecraft efficiently and leave the politicking to others.

After touring Blue Origin's Seattle facility, Jeff invited me to visit his launch site in Texas. Again, I jumped at the chance. *Remote* does not begin to describe Blue's massive swath of real estate in West Texas. My Google Maps went blank as I crossed into the area. I was there for a test flight that ended up being postponed, but I was not disappointed by my tour.

Blue Origin was just completing a new test stand during my visit. Standing at the top with the manager of the project, I asked how much it cost the company to develop. The thirty-year-old Purdue grad thought about it for a minute before giving me his estimate of $30 million. I then asked what I intended to be a rhetorical question: Do you know what it is costing NASA just to refurbish a similarly sized test stand? He quickly answered "$300 million"—the correct answer. He told me he used to work at NASA and had left because of the bureaucracy. I asked another question only somewhat rhetorically: Did he think NASA could test any of our engines at Blue? He shook his head and laughed, saying "Why would we waste our time?"

There was a history behind his answer. NASA has an abundance of people and infrastructure related to engine testing with extensive facilities and capabilities, particularly in Mississippi and Alabama. Test stands that were built in the 1950s and 1960s have since been either mothballed or modified, and more test stands have been built. NASA has spent over a billion dollars on this activity since Apollo and developed only one new engine: the one for the Space Shuttle. A former NASA CFO once told me that of all the nefarious dealings they saw at the Agency, the shenanigans related to test stands was its most likely jailbait.

After NASA canceled Constellation's Ares I rocket's development, the Mississippi delegation forced legislation to complete the test stand for its upper stage anyway, at a cost to the taxpayer of $400 million. As the person blamed for canceling the rocket, I had the honor of spending quality time in Mississippi with the senior senator who made the demands, Senator Thad Cochran. It appeared to me that he cared less about the rocket than about the contractor jobs for the test stand, which was immediately mothballed upon completion and remains a monument to government waste.

Since the country was spending money to keep up the test stands anyway, I thought it made sense to let the private sector—who were actually designing new rockets—use NASA's facilities for their own testing. Partly at my recommendation, Blue Origin negotiated an agreement to test the engines they were developing at NASA facilities, but the two cultures eventually proved incompatible. The disconnect was essentially time. NASA took too much time to do everything. Planning, making decisions, communicating, putting engines on the test stand, and testing were all done at the government's usual timescale. Blue Origin found it was more efficient to build their own test stands in the middle of west Texas, using cement trucks and expertise honed by building swimming pools.

We've all heard the saying that time is money, but the phrase has a double meaning when it comes to procurement. In the commercial world, taking more time to get to market means less money in your pocket. In the government, the more time it takes you to build a spacecraft, the more money they will pay you. The incentives are reversed.

I saw Jeff most recently at a dinner he hosted in DC after his Blue Moon announcement in 2019. He'd gathered about a dozen of us for a relaxed sit-down meal and discussion after his presentation at the DC Convention Center. He held court throughout the feast, asking and answering questions about space and politics—two of my favorite topics. Caroline Kennedy was seated between Jeff and me, and she told us about meeting John Glenn in the Oval Office when she was five years old. After saying hello to the astronaut, she explained, she turned to her father and expressed her disappointment, saying she thought she was going to get to meet the monkey. I'd heard John Glenn tell the same story, but Jeff's loud and infectious laugh signaled he hadn't heard it before.

• • •

I first met Elon Musk in the summer of 2002, when he invited me for breakfast at the Willard Hotel in Washington, DC. He requested the meeting without offering a reason, and all I really knew about him was that he'd recently founded a launch company called Space Exploration Technologies Corp, which later came to be known as SpaceX. We talked about our personal visions for space development and he asked most of the questions. This is what conversations between us have been like ever since. On that early occasion, Elon's questions centered around the Astromom project (which had just ended), my personal interest in traveling to space, and about working with the Russians. We had an engaging discussion, but if it was an interview, I failed.

Elon Musk was born and raised in Pretoria, South Africa. He moved to North America as a teenager, and after starting college in Canada, he transferred to the University of Pennsylvania and earned dual bachelor's degrees in economics and physics. Elon founded a web software company with his brother Kimbal, called Zip2, which was acquired a few years later by Compaq for more than $300 million. He cofounded another digital startup he called X.com, which merged with another company to form PayPal, which was acquired by eBay for $1.5 billion. Elon put his early fortune to good use, starting SpaceX in 2002 and becoming a major investor in Tesla in 2003.

SpaceX was established to reduce the cost of space transportation, something Elon learned was necessary after attempting to find a cheap launch for a small payload he'd designed to go to Mars. Tales of how disrespectfully the Russians treated Elon while he was there to negotiate with them for a launch include how one Russian rocket designer spat on his shoe. This act so completely offended Elon that he decided on the flight home that he would start his own rocket company to compete with them. If Helen of Troy had a face that launched a thousand ships, this was the spit that launched a thousand spaceships.

Intrigued by what I'd heard from Elon, I followed up on our breakfast by visiting SpaceX in El Segundo in 2003. I was working as a consultant, and a colleague and I pitched our services to Gwynne Shotwell, who was the head of SpaceX business development at the time. Gwynne took us on a tour of their new, mostly empty, factory. We did our best to try to convince her that we could provide strategic advice about working with the government, but we never got a contract. I was disappointed at the time, but it was helpful to be able to point out later in my career that I'd never received a dime from the company.

When I met Gwynne next, she was number two at SpaceX, and I was number two at NASA. By then, we'd both earned reputations—not entirely undeserved—as being the people running the show behind the scenes for leading men. We have similar dispositions and have worked together well ever since. I am in awe of what she has accomplished and recognize that, like Ginger Rogers, she has done it all backwards and in heels.

Even though I blew two chances to work for them, SpaceX's goal of reducing the cost of space transportation—untying the Gordian Knot—was my central focus, and I kept tabs on their progress. As the senior space policy advisor to John Kerry's presidential campaign in 2004, it was already obvious to me (and to many others), that the government shouldn't be building and operating its own Shuttle replacement. We were in favor of stimulating private initiatives instead. If Kerry had been elected and I'd come back to NASA in his administration, we may have gotten the policy in place earlier. However, since technical and financial resources were also required for success, Monday morning quarterbacking should be kept to a minimum.

NASA wasn't the first government agency to take a chance on SpaceX. That honor goes to the Defense Advanced Research Projects Agency (DARPA) and the Air Force, which invested a small amount of money, around $8 million in 2003, to demonstrate highly responsive, affordable launch capability with the Falcon 1 rocket. Pete Worden and Jess Sponable were space pirates who spearheaded the support of DARPA and the Air Force and did yeoman's work incentivizing private launch capabilities throughout their government careers.

The NASA program that was developed after SpaceX protested its noncompetitive award to Kistler Aerospace provided the first significant government funding to the company in 2006. The COTS program was designed to incentivize the development of private sector capabilities to deliver commercial cargo to the ISS, and the $278 million awarded to SpaceX allowed them to scale up their rocket and develop their initial cargo capsule—the Dragon. Dragon was to launch on a yet-to-be-designed rocket called the Falcon 9, but the first three test flights of the earliest version of the rocket, called the Falcon 1, failed. After the third failure in August 2008, Elon publicly acknowledged SpaceX was almost out of cash. They only had enough left for barely one more try.

The August failure happened a month after I'd been tapped to lead Obama's NASA transition team, so I'd already conveyed my personal views on the value of using private sector launch services to the candidate. SpaceX was widely believed to be the most likely private company to succeed, so my recommendation wasn't looking too prescient at first. My chips were all on the table, and I wasn't even playing my own hand, so this was not the most comfortable form of gambling.

The COTS model was central to the transition team's early thinking as we assessed options for how to best replace the Shuttle. Although our direction wouldn't have changed if SpaceX's next flight had failed, the successful launch helped convince the Obama policy team to place their bet on starting a Commercial Crew program immediately. President-elect Obama's senior advisors who read our weekly NASA transition reports were supportive of the concept from the beginning, recognizing the potential benefits of privatized astronaut flights. Even so, when the Falcon 1 launch succeeded, it put an exclamation point on that option.

The big prize—contracts to supply cargo to the Space Station—were announced by NASA that December. SpaceX was awarded $1.6 billion to perform twelve Dragon flights, and Orbital Sciences Corporation got $1.9 billion to launch eight flights. It was the beginning of a pattern, where SpaceX signed up to deliver more than their competitors for less money. As of this writing, SpaceX has successfully launched and returned twenty-two cargo missions and Orbital Sciences has conducted thirteen. Adding to the discrepancy is the fact that the Dragon carries cargo up and down, while Orbital Science's Cygnus vehicle is a space capsule with no down-mass capability.

I was working out of my transition team office at NASA Headquarters by the time of the first Commercial Resupply Services (CRS) awards, but I was not involved in the decision. Making such large awards within the final thirty days of a lame-duck administration was typically frowned upon, but my team and I were thrilled with the selection and saw no reason to make a fuss. Elon has often spoken about the importance of receiving NASA's early confidence and funding. Receiving that confidence during a rocky NASA handoff to a new administration was just one more hurdle they overcame.

The fifth (and last) launch of the Falcon 1 in July 2009 was also noteworthy. The launch was the week after our Senate confirmation hearing and one day before our nomination came to a vote on the Senate floor. Again, I held my breath hoping they wouldn't have a problem that would give an opening to those who opposed the government's transition to private sector rockets. Their successful launch put a spring in my step as I walked into Headquarters to take my oath of office two days later.

SpaceX's June 2010 launch was just as critical to our Commercial Crew proposal. This was the first flight of the Falcon 9 rocket that would eventually launch the Dragon capsule, and by now I wasn't the only one with my chips on the table. President Obama had all his in as well. Any misstep at that point would have led to even more criticism of the administration's plan.

Not only did that flight go off without a hitch, but it was followed later in the year by the successful launch of the Dragon capsule into orbit. It carried a giant wheel of cheese to honor Monty Python, and because well, that is what Elon decided.

These were not insignificant milestones, given the ongoing public political debate about NASA's transformational budget proposal earlier in the year. Successful launches didn't end the debate, but failed launches likely would have.

• • •

SpaceX completed their COTS demonstration missions of the Falcon 9 and Dragon capsule the year before the Shuttle retired, giving a ray of hope to those of us who believed they were our best shot at again launching astronauts from the US. In May 2012—a year after the final Shuttle mission—the Dragon became the first commercial spacecraft to dock to the Space Station.

One of the responsibilities I enjoyed at NASA was working with the other international space agencies, and I was in Tokyo during that first Dragon docking. The Japanese Space Agency (JAXA) is proud of its space accomplishments and is one of NASA's most reliable partners. I'd planned to travel to Japan the year before, but the 2011 tsunami and subsequent turmoil in the country caused a postponement. When the scheduled Dragon docking conflicted with my newly rearranged trip to Japan, I knew honoring our partnership was a priority, but was disappointed to have to watch the important milestone from afar.

The mission's launch was on my birthday, giving me two good reasons to celebrate with our Japanese colleagues. My away team managed to get a live feed of the Dragon's docking, and we listened to the operations room activities through a phone line. Our celebration lasted until morning in the land of the rising sun. Before leaving the country, I purchased an ornate dragon sculpture, a symbol of the pride I had in the "Dragon Lady" moniker given to me by people who didn't consider it a compliment.

I wasn't the only person impressed with the successful docking of the Dragon. The day after the flight, the equity evaluation of SpaceX doubled to $2.4 billion. It is valued at more than $100 billion as of this writing.

From a cold start less than twenty years before the publication of this book, SpaceX has become the sixth-largest NASA contractor, employing nearly 10,000 people. Not just NASA but the Air Force and other military and intelligence services are now routine users of their products and services. This is especially notable given how many meetings I sat through in the Pentagon while being questioned and ridiculed for committing NASA resources to the start-up company. Senior industry and government officials took pleasure in deriding the company and Elon in the early years. The people in government who were supposed to be stewards of the taxpayers couldn't believe anyone other than their friends at the United Launch Alliance could succeed. To me, this seemed irresponsible. The exorbitant costs ULA charged were undermining the competitiveness of the US government and industry.

It didn't help that Elon was younger and richer than they were, with a Silicon Valley disrupter mentality and lack of deference toward the traditional industry. Some of the contempt and criticism the status quo displayed against Elon was personal, but anyone with a real sense of what was happening knew their resentment was more fundamental. None of the companies that had previously set out to topple Goliath had come close, but it was a house of cards that couldn't stand for long once SpaceX started succeeding.

People in the government who were reluctant to support SpaceX in the beginning are already busy rewriting history to boast how they were early adopters of the concept. Lest we forget—neither NASA nor the Air Force gave SpaceX contracts until after losing a legal challenge. The Air Force continued to award sole-source contracts to ULA, even after SpaceX was successfully flying NASA missions. In 2014, SpaceX protested an $11 billion sole-source block buy given to ULA, which forced the Air Force into mediation before being bound to open a competition. Even then, the Air Force dragged its feet for years.

COTS similarly only began after SpaceX filed a protest and NASA was told it had to start a competition. It then ignored SpaceX's offer to develop a system to transport astronauts to the Space Station for $300 million, which could have kept us from sending twice that amount to the Russians. Even now, NASA is relentlessly committed to using tens of billions of tax dollars to directly compete against privately funded reusable heavy-lift rockets. Not exactly the stuff of bragging rights.

Over time, NASA employees who were dead set against partnering with the private sector—especially with SpaceX—couldn't help but recognize its ability to outperform the competition. SpaceX so impressed NASA with their progress in meeting the objectives of the COTS program that they were allowed to combine two demonstration missions into one. Their secret was no mystery: they delivered the best value. Time and time again SpaceX was awarded significantly less funding to do more than their competitor and proceeded to outperform them. SpaceX is proof that the theory of faster, better, and cheaper is possible.

A classic example of this was in December 2010, the day before the planned first operational launch of the Dragon. A final pad inspection revealed two small cracks in an engine bell of the Falcon 9 rocket. Everyone at NASA assumed we'd be standing down from the launch for a few weeks. The usual plan would have been to replace the entire engine, which took a month in the Shuttle days. SpaceX did their calculations, assessed their margins, and decided to snip off the end of the nozzle that was cracked. They launched successfully—one day later. COTS funding was through a partnership agreement and not a contract, so NASA couldn't do anything but accept SpaceX's decisions and watch in disbelief.

Elon wasn't the first to recognize the value of reusable rockets, but he was the first to make it economically advantageous. "Reusable" was not even an acknowledged term used to describe previous transportation vehicles when they were developed, since no one would ever have considered making a cart, car, ship, or airplane just for single use. Even so, the aerospace industry standard bearers scoffed at SpaceX's early efforts to test vertical landings of rockets in 2012. For many people, it looked and seemed crazy to try to land boosters returning from space on a barge in the ocean, and videos of their misses and hard landings were sometimes derided behind closed doors. People aren't laughing anymore.

Government efforts to reduce the costs of space transportation through reusability started thirty years earlier on the Space Shuttle program. Several of the Shuttle's parts—including its engines—were designed for reusability, but refurbishment proved to be nearly as expensive and time-consuming as it would have been to build new ones. As usual, the problem was the incentives. Once a company was paid to build engines, they still wanted to employ the same number of people, so costs for testing and refurbishment naturally escalated. No one in the system questioned whether it was a variable that could be modified, since it was easier to keep doing the same thing while charging the government more and more money.

This process continues to this day under SLS development, in which NASA is paying Aerojet Rocketdyne $150 million per engine to "refurbish" Shuttle-era engines they had already paid them to build, that were sitting in a warehouse. Since the SLS throws four away each launch, taxpayers will spend $600 million per launch for engines they paid for already. By contrast, SpaceX sells a Falcon Heavy launch for $90 million, reusable engines included.

This is as good a place as any to recognize that publicly traded companies are beholden to shareholders, who tend to value strong quarterly reports, increasing stock prices, and dividends. Aerospace companies' leadership focuses on maximizing short term shareholder value, which includes taking advantage of government incentives, as well as loopholes and gray areas. It is the government's responsibility to establish and enforce policies that award innovation and efficiency, both in Congress and in the administration. It is in the nation's best interest to have a competitive industry, so sticking with policies that reward laggard capabilities and overlook ethical or business infractions hurts our economy and national security.

As a government employee, my job was to advance and improve US aeronautics and space capabilities to deliver greater value to the taxpayer. Defending Elon and SpaceX was never my job or mission. I wasn't in the tank for them; I would have been supportive of any private company taking such bold initiatives to improve US competitiveness. False rumors about various nefarious links between us were circulated during my tenure. A more serious concern should have been raised about why other government leaders were being so critical. What were their reasons for denigrating the person and company responsible for saving billions of taxpayer dollars and making the United States competitive in the world launch market? People in the space club gossip about both Elon and Jeff's divorces, forgetting how many first wives their astronaut and industry CEO friends have left behind.

Even though I've never worked for any of these guys, I can still get defensive about the double standard often displayed both in and out of the space community.

My story is difficult to separate from Elon's because I wouldn't have managed to pull off much of a transformation at NASA without him and SpaceX. We've bled for the same cause and amassed some of the same enemies. In a 2012 *Esquire* article about Elon titled "Triumph of His Will," author Tom Junod wrote, "He maintains what an official with close ties to NASA calls a 'symbiotic' relationship with Lori Garver." We each needed the other to succeed.

At the time we put the competition in place, I was confident that more than one company would be able to offer astronaut transportation services. Similarly, Elon likely had confidence that many senior government leaders would see the value of using private companies to launch astronauts to the Space Station. Both circumstances would eventually become true, but they weren't at the time, and delays on either of our sides could have led to an entirely different government strategy.

Like most geniuses I've known, Elon doesn't mince or waste words. He asks questions and listens to your response before reacting, unless or until he decides you are an idiot. His mind works quickly, and he's not one for small talk—at least not with me. The last one-on-one meal I shared with Elon was scheduled as drinks in 2012. We'd gotten to know each other a bit by this time, but I worked more closely with Gwynne Shotwell and Elon's government affairs team. SpaceX staff asked if I was available to meet Elon after work before he had to get on a flight back to California, and I was more than happy to oblige.

The intensity of our roles and mutual challenges we faced made it especially nice to unwind and blow off steam. We ordered tapas, a pitcher of margaritas and talked for several hours. I'm not sure either of us were accustom to consuming that much alcohol. Looking back, it is surprising that no one recognized him or interrupted us. That certainly wouldn't happen today. When I finally looked at my phone, I saw his staff had been trying to reach me to remind him that his jet was fueled and waiting for him to depart.

As of this writing, I haven't talked with Elon in a few years, and our most recent direct communication was a dust-up on Twitter. He misinterpreted a poorly worded quote of mine about the importance of the business case for launching satellites and responded that he could have more easily made more money starting internet businesses. My message was meant to highlight the value of opening private sector markets, not insinuate that he was only in the business to make money. I'd assumed my willingness to take heavy fire to advance the policies required for SpaceX's success would be enough not to question my motives. But thousands of Elon's followers took offense, which led Elon to caveat his original tweet. I've always found our in person conversations open and unpretentious. An example of this can be seen in a discussion that included my oldest son and is one of my favorite stories about them both.

It was June 2014, and Wesley had just returned to DC after graduating from college. SpaceX hosted an event at the Newseum to unveil their Dragon V2—what became Crew Dragon—and I brought Wes as my guest. I hadn't seen Elon since leaving NASA, and when he asked me about my new job working for the airline pilots' union, I told him—only half joking—that I hoped to include his pilots in the union someday. He laughed and said there was no need for pilots on Dragon and then turned to Wes and asked what he had majored in at college.

As soon as Wes said his degree was in music composition, Elon responded that it was another field that would soon be entirely automated. I thought it was a rude thing to say to a young college grad, but Wes didn't appear offended and didn't hesitate to disagree with the imposing icon.

Elon countered Wes, point-for-point, saying that even the creative flair and imperfections that composers were known for could be written into the software. Accepting Elon's points, Wes offered that as listeners, the meaning and feelings evoked by many compositions was tied to our knowledge of the individual composer. He then asked Elon if a Bob Dylan song would be the same if we didn't know Dylan had been the composer. After a short pause, Elon nodded his head in agreement and said, "You know, I think you are right."

This is a favorite story about my son because I am proud of his confidence and ability to convey such meaningful insight in the face of being challenged by someone famous for being brilliant. It is also a favorite story about Elon. Elon is known for being brash and unconcerned over hurting people's feelings. He didn't know my son, knew he had gotten his degree the week before, and had no problem saying his chosen field was a dead-end career. But he also listened to a twenty-one-year-old, thought about what had been said, and changed his mind. As an observer of the conversation, I was proud of them both.

• • •

In addition to developing transportation systems for the Space Station, SpaceX started investing its own money to develop a larger, partially reusable rocket called the Falcon Heavy in 2011. Even after the Falcon 9 had been successful, Elon's critics said he would never be able to build something to launch heavier payloads.

I happened to be at NASA's Marshall Space Flight Center (MSFC) in April 2011 when Elon announced he was going to build the Falcon Heavy. We were still in the throes of the battle over the design of NASA's planned heavy-lift booster, and the rocket boys in Huntsville were not pleased with the announcement. NASA's leadership at the Center made what they seemed to think was a reasonable request, asking me to tell Elon to stick with the smaller rockets. Big rockets were in their lane. I was glad to be meeting with them in person, so I could more fully explain why their request was ridiculous.

These were rocket scientists, not political scientists, so perhaps this was new territory. I described why the government's relationship with US industry wasn't a competition. We weren't racing against them in a different lane. I suggested they think about it in terms of a cycling peloton, and it was our job to ride out front so our commercial team members could draft behind us. If a company in the pack gained the strength to pass us, we weren't supposed to grab our tire pump and stick it in their spokes; we needed to wave them past and look for a new hill to climb. I thought it was a great analogy, but I'm not sure competitive cycling is that big in the South.

Elon has been public about the fact that the Falcon Heavy was a bigger hill to climb than SpaceX initially predicted. It took them longer and cost them more, but they weren't spending tax dollars, so there wasn't any reason to be critical. The first launch was finally scheduled for February 2018. I'd already left NASA, but SpaceX sent me an invitation, and I made the trip to Florida to see it in person. SpaceX had rented the usual VIP viewing site from NASA and it was a perfect view of the Falcon Heavy standing majestically on the same pad where the Mercury, Gemini, Apollo, and Shuttle missions had launched—the pad they had competed to lease when I was deputy.

The rocket is similar to the Falcon 9 but has additional side-mounted boosters designed to return to the launch site under their own power for reuse. SpaceX was regularly landing its single boosters on a barge at sea, but boosters returning to the Cape made for tighter local wind and weather conditions for launch. High winds delayed the launch for several hours as the group of about one hundred well-wishers mingled nervously on the balcony.

I'd talked with one of my senior SpaceX former colleagues at the NASA holiday party the year before, who told me they'd offered to launch a government payload on the first flight of the Falcon Heavy at a deep discount. He said they were turned down because of the risk of flying on an untested rocket. NASA has specific payloads available for similar test flights, such as student payloads, but for one reason or another, seems to have demurred. When Elon announced that he was launching his personal Tesla Roadster, there was no mention of any earlier offer to the government. Critics called his payload choice superfluous.

Being back at the Kennedy Space Center for a non-NASA SpaceX launch was an alluring experience, and I was happy for the extra time while we waited on the weather. The majority of VIPs in attendance were finally space pirates and the fate of the launch felt determinative to how the bounty would be divided in the future. We were already discussing changing our plans to come back the next day for another launch attempt when the winds abated and the countdown resumed. The surprise announcement was greeted with excitement as we made our way to the railing to watch the maiden voyage of the biggest rocket to launch since Apollo. Three, two, one...

The intensity of the crowd's emotions piqued as the Falcon Heavy's twenty-seven engines cracked to life simultaneously, lighting up the sky like a second sun as it slowly lifted into space. As with watching all launches from a distance, you see the rocket launch several seconds before you hear or feel the vibrations, since light travels faster than sound. I'd seen a lot of Space Shuttle launches from the same vantage, and it was always thrilling. But this was louder, and the sound waves that reverberated in my chest were even stronger. People who'd watched both Saturn V and Shuttle launches said the Falcon Heavy launch effect was closer to that of the Saturn V.

Still on the balcony and teary-eyed from the elation over the launch, I wasn't prepared for the two distinct sonic booms signaling that the rocket's boosters had done their job and were headed back our way at subsonic speeds for their return to their landing site across the Banana River. Appearing at first like entering missiles then simultaneously spinning and slowing for their choreographed landing, the boosters gracefully set themselves down on their tails next to each other before shutting off their engines—like a pair of Olympic divers entering the water together, precisely straight with no splash, a perfect ten.

The viewing parties' focus quickly shifted back inside to the large video screens showing the rocket's nose cone protecting the payload slowly opening in space, revealing Elon's cherry red Tesla Roadster being "driven" by a mannequin in a space suit. Three cameras had been attached to the car to show different views of the most creative and surreal spacecraft of all time.

We watched in awe as the Tesla drove past Earth on its way to the Red Planet, while listening to the playlist Elon had preset in the car's radio, starting with David Bowie's "Starman." Telescopes have already watched the little red Roadster speed past Mars' orbit on its way out toward the asteroid belt. I penned an op-ed that ran in *The Hill* the next day calling it "cross-marketing genius," which some viewed as a criticism, but it was meant sincerely.

The focus of my op-ed was a call for NASA—my former employer at the time—to end its fixation with building its own big rocket at great expense to the taxpayers now that the Falcon Heavy had proven itself. I'd recently seen a brochure for the Space Launch System that NASA was distributing at a conference, showing how it could launch 12.5 elephants to low Earth orbit. The colorful marketing piece had pictures of elephants stacked neatly in the rocket's cargo hold, one by one, trunk to tail. This was the type of thing that drove *me* batshit crazy. I did a back-of-the-envelope calculation to compare how many elephants the Falcon Heavy could launch—9.7—and used it to explain the absurdity of NASA's plan.

I pointed out in the op-ed that using tens of billions of dollars of the public's money to build a rocket that if/when it was completed could launch 2.8 more elephants than a rocket already developed and launched at no cost to the taxpayer was wasteful. Even setting aside the $15 billion (now $20 billion) sunk cost to build NASA's rocket, the comparative per-flight cost of each would allow the Falcon Heavy to launch 84 more elephants for the same price as the SLS. Impressive indeed!

The size of a rocket is typically driven by the necessity of what it is designed to launch—its payload requirements. The Falcon Heavy had been sized to launch very large and expensive military satellites, as it has done several times already, for about $150 million each. Marketing the SLS to the public (and spending the public's own money to do it) by showing it could launch a bunch of elephants exposed the fundamental problem. Building a really big rocket was their purpose, no other justification needed. Launching Elon's Tesla Roadster seemed positively reasonable by comparison.

Completely disrupting the global launch market by reducing prices and increasing reliability has already made SpaceX the most sought-after provider of rocket launches both in and out of the government. Still seen as disrupters, SpaceX has single-handedly returned the United

States to a leadership position in today's space race. After launching close to zero commercial satellites twenty years ago, the United States launched more rockets to orbit than any other nation in 2020. SpaceX conducted twenty-five of the launches, compared to six for the United Launch Alliance. New start-up companies in the United States launched an additional nine, giving the US forty launches compared to thirty-five for China and seventeen for Russia. All other countries remain in the single digits.

Not only has SpaceX completely reversed the US strategic and economic position related to rocket launches, but the company is also building many of the payloads.

SpaceX announced in 2015 that they were developing their own satellite internet constellation called Starlink. Moving at what seems like a lightning pace even for SpaceX, there are already more than 2,000 satellites operating on orbit and thousands more coming soon. The system is, like all things SpaceX, disruptive and controversial.

Elon's vision for space development is similar to many other space pirates, to make humanity a multi-planet species. His chosen planet for our first additional home is Mars, and he is already developing a system to go there—he calls it Starship. Elon's vision for Starship is to carry a hundred people at a time to Mars, with a goal of populating Mars with one million people by 2050. That is not a typo. He's already begun testing various stages of the reusable rocket at a fast-growing facility in East Texas that he calls Starbase.

The policies and programs that had been devised for over a decade were created to provide breadcrumbs for the private sector to follow. Although our work was necessary to eventual success, it was never going to be sufficient. What Elon and his team have already accomplished at SpaceX is the transformative force. The doctrine of how it could be done existed in many minds earlier, but SpaceX made it so. Considering that this is only one of the industries Elon has disrupted is beyond my own comprehension.

def. The process of moving, changing position or changing place; orientation of a body over time; translation

9.

IT'S NOT JUST ROCKET SCIENCE

ROCKET SCIENCE IS TYPICALLY SEEN AS FAR MORE DIFFICULT THAN POLITICAL science, but the opposite has often proven to be the case. The intricate design, manufacturing, and operations of launching something into space is exceedingly more complex than our system of government, but gravity is constant and consistent. Even though it is an incredible challenge to overcome, especially when lifting something heavy, smart, trained people who follow the laws of physics and work together can overcome it. The same people have found it harder to work together to adhere to the laws of politics, which has left the human spaceflight program continually chasing its tail.

The mandate to lower the cost of space transportation set in 1970 by President Nixon was to reprioritize this "massive concentration" of energy given to NASA, in favor of developing "low cost, flexible, long-lived, highly reliable, operational space systems with a high degree of commonality and reusability." Just imagine what could have been accomplished by now if NASA had accepted and successfully delivered on this mandate as it had its previous goal to beat the Russians to the Moon?

Instead, NASA leadership designed the spacecraft they wanted to build, prioritizing internal interests and parochial constituencies instead of embracing the assigned national political mandate. In my view, we have been working the wrong end of the problem—putting the cart before the horse. Programs that require large amounts of taxpayer funding must be guided by established valuable purposes. The cart must follow the horse. Building a big rocket or going to a specific destination aren't ends in themselves; they are means to an end.

Neil deGrasse Tyson has referred to the space community's fixation with repeating similar types of programs as "Apollo necrophilia." It is time to accept that the national purpose that drove our first missions to the Moon have long since passed. Neil has observed that throughout history, the most significant public expenditures are tied to at least one of three motivations—fear, greed, or glory. In the case of Apollo, we feared the Soviet Union gaining global strength through their successful space exploits, we gained economic benefits by investing in new technologies, and accomplishing the goal was glorious. Other historical examples Neil offers to back up his theory include the building of the pyramids and the Great Wall, and Queen Isabella's investment in ships to find new trade routes and show her country's strength.

I subscribe to Neil's doctrine, and like many NASA supporters, believe at its best, human spaceflight programs can contribute to meaningful purposes. But the NASA community too often fixates on building the cart we want to build without considering the motivation of the horse. You can yell, whip, and try to pull a horse, but if they don't want to move—or your cart is too heavy or has square wheels—you aren't going to get very far.

Constancy of purpose has recently become a catch phrase used try to keep existing programs from being canceled by new administrations. I recently addressed this topic during a talk I gave to NASA project leaders who expressed concern over politically driven policy changes. I suggested they look at the problem differently. Since they were engineers, I reminded them that as long as we are a publicly funded agency, the democratic system is the constant and NASA programs are the variable, not vice versa. Imagine if a kid who regularly overspent his allowance on candy, blamed running out of money and getting cavities on his parents. Unless we want to emancipate ourselves from the government entirely, the best way to earn our allowance, is to design achievable programs that align with our given national purposes, and then deliver them as promised. Consistently. Every parents dream.

NASA's *purpose* already has constancy, since it is derived from the NASA Space Act. The debate isn't actually over our purpose—it's about how to best achieve that purpose.

The widely accepted purpose for human spaceflight over the last sixty years has centered on providing public inspiration, national economic growth, and some form of international leadership—either through competition or cooperation. These are variations on fear, greed, and glory. Human space exploration hasn't gotten very far partly because we haven't been designing programs to best deliver on these values since Apollo. Just as saying the sky is purple doesn't make it so, saying what you are doing is inspirational, stimulating the economy, and providing global leadership doesn't make it so either. Like most children, NASA has more control over our parents than we realize.

If we want to spend the people's treasure to "inspire" them, we need to consider whether what we're doing is truly inspirational enough to justify the cost. If we say programs will provide jobs and stimulate the economy, but we fall back on government contracts that don't innovate, drive new technology, or leverage new markets, are they really returning the best economic value? If we decide what we want to do and how to do it, and only then offer other countries the opportunity to join us, or try to fabricate the same race against a new enemy, have we maximized our global leadership position?

When I joined the transition team for President-elect Obama, I shared an office in the temporary White House set-up for the incoming administration with the person who led the transition for the Office of Science and Technology Policy. Our hall was populated with the leads for the National Science Foundation, the National Oceanic and Atmospheric Administration, National Institutes for Health, and so on. NASA was part of a group referred to as STARS—Science, Technology and the Arts—led by Tom Wheeler, who later became the Federal Communications Commission Chair. The team that ultimately filled senior science and technology positions in the administration had an aligned purpose to advance economic and social benefits. Not all of them were too excited about NASA and human spaceflight at first, primarily because it wasn't clear how what we were doing contributed much public value. I believed that NASA's programs had the potential to contribute to greater economic and societal benefits if they were restructured, and knew we had a short window to make that happen.

Earth sciences and aeronautics programs were the most naturally aligned with these purposes and therefore received a high priority in our 2009 stimulus budget request. The increased funds weren't provided to build more of the same carts, but to drive technology and innovation: to deliver the value that motivated the horse. One obvious way to improve value is to lower the cost and time involved in building the cart, which then gives any horse a better chance at pulling it successfully. Working this side of the problem was a primary driver for the NASA policies and programs I tried to establish in the Obama administration. An obvious way to sustain and expand space activities is to liberate NASA from its crushing infrastructure and transportation costs.

Traditionalists at NASA and in industry accused the administration of wanting change for change's sake. In truth, change was required to deliver programs that could better fulfill the purposes for which they were established and promised. Driving to a consensus among the leadership over how to assure NASA's programs could be more relevant and sustainable was challenging from the deputy position. Without NASA's leadership aligning with the administration's vision, progress would be unattainable. Sean O'Keefe had written a vision statement for the Agency that still appeared on NASA documents. It read, "To improve life here, To extend life to there, To find life beyond." Some of us referred to it as the Dr. Seuss vision.

Early in our tenure, at my request, Charlie agreed to hold one of our quarterly senior leadership meetings offsite so the team could focus on creating a new vision statement to better articulate our purpose. I recruited a world-renowned expert in helping organizations reach their aligned vision—Simon Sinek—and carved out a few hours on the agenda. Simon first asked the team to write down our view of NASA's "best day." I expected it to be the Moon landing, but a consensus formed around saving Apollo 13. This finding was an *ah-ha* moment that led to a shared team view that a part of NASA's DNA was to rise to new challenges, seeking to know things that were previously unknown.

The team generated an initial statement within a few hours: *To Reach for New Heights and Reveal the Unknown*. It was good, but Simon pushed us further...Why do you do those things? It was an important conversation and I remember a few of the cup boys struggling with the concept. What do you mean *why*? We do it to go to the Moon or to go to Mars. "But why?" Simon asked. "What is the purpose of your discoveries and what are the results of going farther? Who are your customers and how does what you do benefit them?" The discussion generated another clause: *so that what we do and learn will benefit all humankind.* The statement was the result of a collective effort by the leadership team, and I remember it being a rewarding exercise.

Per our employee-bargaining agreement, a union representative was allowed to be present for management decisions, and I had begun the practice of including their elected leader in our leadership's quarterly meetings. Charlie and a few others in management didn't agree with the practice, but this was one of the areas where, because of his initial missteps, senior White House officials had insisted I take the lead. The President of NASA's largest employee union attended the retreat but in no way dominated the discussion.

The vision statement withstood the test of time. More than twelve years, three administrations and six strategic plans later, NASA's statement carries the same message, with improved efficiency: *To Discover and Expand Knowledge for the Benefit of Humanity*. Words matter, and I am still proud of helping the Agency reach alignment on such a meaningful NASA "why" statement.

The Administrator's best mate and cup boy Mike Coats was at the leadership retreat, but came away with a different view. Mike complained about the statement in an interview after he retired saying, "The Obama administration came in. They weren't interested in space, so how can the space program help him be reelected? How can it help the Democratic Party? How can it help the unions? For the first time, the union representatives sat in on all management meetings at NASA. Very vocal. Read the mission statement for NASA. That was written by the union representative. Makes no sense to me. It could have been written for McDonald's french fries, for all you know. It had nothing to do with space. It doesn't mention space in there at all. It was literally written by the union representative, and Lori insisted that they adopt it."

Arguing that the statement, "To reach for new heights and reveal the unknown, so that what we do and learn will benefit humankind," could have been written to sell McDonald's french fries is preposterous. Mike's complaint that our mission statement was "written by the unions" is also untruthful and confirms how uncomfortable he was with their participation. So much for caring about the workforce. I visited JSC numerous times while Mike was the Center director and held all-hands meetings with just the two of us on stage. These were not easy conversations, but I hadn't understood the extent of his personal resentment until I came across the public "oral history" interviews he conducted at JSC.

In one of his more revealing rants, Mike told the interviewer, "It's not unusual to have the Deputy Administrator be a political type ... but she wanted to get involved in the technical decisions, in the management decisions. Remember, Lori had no executive or management experience. None, zero, zip. And she had no technical background. She prided herself on not being technical, and now she's the Deputy Administrator of NASA. She wanted to fix everything right off the bat, and really not much was broken, at least on the human spaceflight side. Because she had no management experience or executive experience, she really didn't have much to offer to help, and she didn't even know the right questions to ask."

The reforms I was advancing ran headlong into Mike's and other cup boys' world views. It seemed he could not imagine a person like me adding any value to the human spaceflight program, which he didn't view as in any way broken. To Mike and many of the cup boys, I would always be a square peg who didn't belong. Mike Coats and others who were personally and financially invested in NASA's legacy programs were a part of the system that lost two Shuttles and failed to develop a realistic follow-on program. They were the team that had gotten us into the current hole in human spaceflight, so it wasn't surprising that their solution was to just keep digging.

Alternatives to the Space Shuttle, which had been proposed since the 1990s, should have led to operational reusable space transportation systems before its planned retirement in 2010. The X-33 program of the late '90s became the Space Launch Initiative in 2001, which led to the

Orbital Space Plane (OSP) in 2003 and Constellation in 2005. All but X-33 were designed to be owned and operated by NASA.

Dramatic, immediate changes were required to put the human spaceflight program on a stable and sustainable path by 2008. I believed our best chance to shorten the inevitable gap in human spaceflight was through a competitive partner-type program. Requesting funding for COTS-D in the early 2009 stimulus bill was a risky move, but I thought the prospect of the significantly accelerated development of a Dragon certified to carry astronauts was worth the risk. The controversial action brought haters out for me early, but it also helped us jam our foot in the door, giving us the opportunity to pry it open later. Even though we didn't get the entire amount requested, the small team assigned to work on the project put the $90 million we received to good use.

Establishing a small program office and initiating commercial agreements with private sector partners didn't keep traditionalists in the NASA bureaucracy, much less Congress, from opposing the development of a sustained program. In their view, we'd gotten the stimulus money on "a technicality" and they had no intention of codifying a way for the private sector to launch astronauts. The very idea was a threat to tens of billions of dollars in Constellation contracts, and they held most of the cards. We'd won an early hand, but the game was far from over and the people on the other side had been running the table for years.

The NASA bureaucracy didn't request money for commercial crew activities in either of their next two annual budget submissions. Chris Scolese oversaw the 2010 process, since that budget was submitted during the months-long standoff between Senator Nelson and the President over who should run NASA. The best we could manage without new leadership was to submit a placeholder budget for human exploration, noting that it would be reevaluated after receiving the report of the presidentially established review committee.

When Charlie disregarded the President's priorities in the next budget cycle, fiscal year 2011, I was forced to make a difficult choice. Others may have chosen differently. I facilitated the work behind the scenes of a small NASA and EOP team to develop and scope a Commercial Crew program using Space Act Agreements. The activity was not supported by or known to the Administrator, but I'd been open with Charlie about why I believed it was neccessary. Unlike the plan NASA put forward that he'd endorsed, it followed the administration's guidance.

Members of the team had different responsibilities for filling in the details of the President's direction. Estimating the Commercial Crew budget was the responsibility of Rich Leshner, a NASA employee on loan to OSTP who had spent years developing the exploration program budget at NASA Headquarters. His work informed the President's proposal of a $6 billion development program over five years, which would fund at least two competitors. The projected funding profile estimated initial flights in 2016, which, if Congress had approved, SpaceX would have come closer to meeting and could have left a shorter gap to fill with purchased Soyuz seats.

Congress made another attempt to reject NASA's development of the Commercial Crew program by including language in their final Fiscal Year 2011 Appropriations Bill eliminating all "new starts." The appropriations language superseded the authorization language, so the Administrator, general counsel, and other NASA leaders accepted it as their final direction. Their "oh well, we tried" reaction reminded me of the scene in the movie *A Christmas Story* when Flick gets double-dog-dared to put his tongue on an icy flagpole during recess. When the bell rings signaling recess has ended, and he's still stuck to the flagpole, the rest of the kids just head back to class, and, as poor Flick squeals, Ralphie looks back to explain with a shrug, "The bell rang."

The prevailing view at NASA was that the bell had rung, but not all of us were ready to let human spaceflight get stuck to a frozen flagpole without challenging the assumption. The Hill's and NASA's initial view that the program was a new start didn't seem credible, so the CFO and I looked for a second opinion. We turned to one of the more creative members of the legal staff, Andrew Falcon, who concluded it didn't fit that definition since NASA had already offered industry the opportunity to fly astronauts through the COTS-D program, and funding for Commercial Crew had already begun through the stimulus budget. We would have lost at least another year if we'd lost the argument. Creative lawyers can be space pirates, too.

The program executive for the Commercial Crew program was Phil McAlister, and I had total confidence in his leadership. He is an unsung space pirate hero who deserves a significant amount of credit for the program's success. One of my primary goals was to make sure he got the people and resources required to facilitate his effort. Trying to manage the program was like a giant game of whac-a-mole. Just when we thought we had one problem solved, three more would pop up in different areas. The bureaucratic battles we fought throughout the first years of the program were numerous and included budget, safety, procurement strategy, personnel, and a commitment to fair competition. It was exhausting.

At one level, I wasn't surprised by the overwhelmingly negative reaction to incentivizing the private sector to build on the commercial cargo program within NASA. COTS benefited from being a low-dollar, less-threatening activity, but Commercial Crew struck at the heart of NASA's culture—human spaceflight. Its budget request was proposed at the same time the large, and therefore popular, Constellation program was being canceled. Charlie and key members of Congress like Senator Nelson eventually got on board, but they were two of its most formidable opponents in the beginning.

Mike Griffin made it clear from the start that he had no intention of extending the program to flying astronauts. He deserves credit—along with OMB—for shaping the commercial cargo program when GAO forced NASA to establish a competition, but he consistently expressed his opposition to expanding the program to include crew. Once it began he said, "Even if they do succeed, that is not a reason for the government not to have its own capabilities. I find it bad policy to put the US government in a position where it is hostage to the services of commercial contractors, with no government alternative. I'm sure that the commercial contractors like that situation. I just consider it bad policy."

Arguing it is better "policy" for the government to subsidize outdated programs that take over a decade to develop, cost the taxpayers tens of billions of dollars, limit the government to only one US alternative, and stifle competition, economic expansion, national security, innovation, and progress is ludicrous. No objective analysis would find that to be "good policy" in my view.

Charlie has publicly confirmed his initial lack of support for the Commercial Crew concept on numerous occasions. His inability to defend or articulate the value of the program in the early years gave the opening for Congress to direct NASA to reinstate Orion, build the SLS, and starve the Commercial Crew program of funding. His standard line is that he "eventually overcame [his] early concerns." His remarks at my own farewell party in 2013 were quoted in the trade press, saying he was "not a 'believer' about Commercial Crew at first, but she had made a difference in his own attitude and that of others. She 'persisted,' he said, adding you've got to 'give Lori credit.'" More recently Charlie has even gone out of his way to brag about his early opposition to the program.

In a televised interview in late 2021, Charlie said that he was "an extreme skeptic" initially. He said, "I went from being the President's selection to be the NASA Administrator, to being probably one of the most despised people in the President's orbit because I did not fall in line; I did not fall in love with the concept of commercial space." Charlie now appears proud of his lack of support for his own commander in chief's policy priorities, adding, "I was not an ideologue like many around me who felt that all we need to do is take NASA's budget, take everything for human spaceflight, and give it to Elon Musk and SpaceX."

Referring to me and others in the administration as "ideologues" for wanting to invest in private sector innovations that would lower the cost of access to space may be meant as derogatory, but in reality, had been directed by policy for decades and was 100 percent aligned with the President's views. Characterizing our request of less than 5 percent of NASA's budget to hold a competition for commercial companies as, "giving human spaceflight to Elon Musk and SpaceX" is hyperbolic and incendiary. It is a reminder of what we were up against in developing the program. Creating and supporting a program Charlie opposed fractured our relationship, but he left me no alternative other than to oppose the administration we had both signed on to serve.

Consistency is not his strong suit. In a late 2016 interview, Charlie said, "We never would have gotten acceptance of Commercial Crew at NASA without my championing it." Perhaps true in his later years, but in my view, this is akin to a lifeguard taking credit for saving someone's life after withholding their life preserver until the victim has struggled to reach shallow water. It's nice he eventually got his feet wet, but his initial instinct was to keep it underwater and his actions nearly sunk the program. Charlie's early derision and doubt added nearly insurmountable weight to the task of implementing the President's highest priority program from the deputy position. His pleasant disposition and popularity cast my own support of the program as the deviation. Post-game pronouncements aside, if either NASA Administrator Griffin or Bolden's views had prevailed, the Commercial Crew program would not have been established.

Every new round of funding brought new criticisms from detractors inside NASA Headquarters and at the Centers, who made attempts to stall progress. The group of men the Administrator listened to above all others he called the "technical authority." The leader of this group was Bryan O'Connor, the third astronaut and midshipman who had graduated from Annapolis with Charlie and Mike in 1968. Bryan was head of NASA's Office of Safety and Mission Assurance, a large and influential organization that favored traditional contracts. The chief engineer and chief medical officer joined Bryan in the troika that Charlie referred to as his conscience.

No one in the "technical authority" chain supported using private partnerships for human spaceflight. Period. They constantly questioned the decision, and the administration's right to make it. They wanted the government to own and operate the systems that carried astronauts in perpetuity, and tried to contort my different view as entirely motivated by politics and therefore out of order.

NASA's Aerospace Safety Advisory Panel (ASAP)—an external advisory committee—was also fundamentally opposed to the program. The amount of negative attention and reviews the ASAP gave Commercial Crew compared to the human spaceflight programs (which were receiving five times the amount of NASA's budget) was absurdly disproportionate. Like the internal "technical authority," the ASAP argued philosophical points against public-private partnerships, and I questioned how they inherently undermined safety. When pushed to the brink, I reminded them that government owned and operated systems were obviously not inherently safe—referencing the Challenger and Columbia accidents. Those weren't easy discussions, and raising the point didn't make me popular, but I spoke the truth. They shouldn't have needed reminding.

NASA's tendency to add layers of bureaucracy and managers to its organization, as though it were a typical FAR-based program, caused more strife. I'd hear about dozens of people being assigned to the program only after it was too late to make changes. NASA assigned employees to colocate with the commercial partners at a rate that began slowing their progress. I had some success limiting staffing levels and making personnel changes, but the bureaucracy did everything they could to push back, and I was often overruled by the Administrator.

• • •

NASA's first partnership awards for Commercial Crew were made in February 2010. We split $50 million of the stimulus budget among five companies. These early funds focused on developing specific milestones that were negotiated separately with each partner. Blue Origin won initial awards, but decided not to compete later, preferring to continue its spacecraft with the private funds of Jeff Bezos. SpaceX lost their first-round bid but persevered in the next round of the competition, which was awarded in April 2011. Four companies won second-round agreements, with awards totaling $269 million.

The next phase was the competition to develop systems that could ultimately be certified by NASA to carry astronauts to the ISS. Phil McAlister and his team, as well as myself and our White House leadership, assumed the third round would proceed again with Space Act Agreements, followed by purchase of services through fixed-price contracts, as was being done in the COTS program.

This plan ran into a buzz saw in 2011. Along with the technical authority, the program office and lawyers preferred to immediately transition to FAR-based contracts. They wanted control. I couldn't keep Charlie from signing off on the plan, and Phil McAlister was directed to work on FAR-based fixed-price contracts. The team held an internal Program Strategy Meeting in July, where the plan was approved. NASA held an Industry

Day in September and released a Draft RFP later that month. Transitioning to FAR-based contracts at this point in the program would allow the NASA bureaucracy to take back control, change requirements, increase costs, and stall progress. I was disappointed and defeated.

Those of us in favor of partnership agreements needed a miracle—like President Clinton's call to Barbara Mikulski the night before Jim Lovell's medal ceremony. We got one. At the last minute—after the bell had rung—Congress unwittingly helped us put the pieces back together for a successful program.

NASA requested $850 million for the program for 2012, and Congress gave us less than half—$406 million. Even NASA's senior management had to acknowledge it was not enough to go forward with the strategy to award two fixed-priced FAR contracts. Being sent back to the drawing-board to consider options was just the break we needed. The team considered stretching the current program, as well as selecting one fixed-price contractor. I pressed to reopen the Space Act Agreement option, and at the last minute Phil was given approval to add it to the chart in the final briefing package.

Bill Gerstenmaier, known as Gerst, was the head of NASA's Human Spaceflight Directorate. He and Phil presented the options to Charlie and me in a briefing that December. Gerst recommended down-selecting to one fixed-price contract immediately, with the potential for exercising the SpaceX COTS-D option in the future. This scenario all but assured Boeing would be awarded the contract. Charlie said he would let them know his decision the next morning.

When assessing the relationship between Charlie and me, some observers have characterized him as being my puppet. This was not the case. It was most often Charlie's cup boys pulling his strings. For much of my tenure as deputy, I felt like Lyndon Johnson did about the Vice Presidency: it wasn't worth a warm bucket of spit. Hundreds of NASA leaders had more budget authority than I did. Heck, I didn't even have the authority to make a $5 million decision for an existing green fuel aeronautics project without getting run over. In situations like mine, where no specific authorities are delineated by the first chair, the power of a second-chair position is often derived from access and persuasion.

The night of our meeting with Gerst and Phil, I knew I had to choose my words carefully and be the last person in the room.

I made the case to Charlie that although proceeding with another partnership agreement was not the internal team's recommendation, it was the only acceptable path for all parties. I first observed that the NASA team wouldn't have offered it as an option if they didn't think it deserved reconsideration, and then I suggested to him that the Hill's action had virtually eliminated the feasibility of other strategies. He didn't need much reminding by then that it was the only option consistent with our guidance from the White House. I closed by sharing my view that if he and NASA embraced the plan as their own, the Hill was likely to buy in too.

When Charlie told the team the next day that he had decided to go with a Space Act Agreement partnership option, Gerst looked about as disgusted as I'd ever seen him. It was a critical decision and allowed the companies to make another two years' worth of progress before the major NASA landing party would come with a FAR-based contract. It is by FAR the most consequential thing I did for the Commercial Crew program during my tenure.

We called the last partnership round Commercial Crew integrated Capability (CCiCap). It financially supported the development of proposals through the middle of 2014, when the decision would be made about which companies would move forward for certification. NASA selected three CCiCap winners in August of 2012: Boeing for $460 million, SpaceX for $440 million, and $212.5 million to Sierra Nevada.

After agreeing to support a commercial crew program in exchange for the administration's supporting SLS and Orion, Congress gave the program lip service while cutting the appropriations request nearly 40 percent over its first four years. NASA's Congressional Appropriators transferred hundreds of millions of dollars from the requested Commercial Crew budget to the billions already provided to SLS and Orion. Instead of receiving the budgeted $6 billion over the first five years, the program received $4.2 billion. During the same period of time, NASA requested $15 billion for SLS, Orion and their ground systems, but the programs were appropriated even more—$20 billion from Congress.

SpaceX had a leg up in the CCiCap competition, since they were meeting their milestones on commercial cargo. But this also meant that any significant problems they encountered would also be highly visible to NASA decision makers. The evolutionary nature of the SpaceX system was a compelling differentiator—as they gained more experience launching satellites, the rocket's reliability would increase, and costs would decrease. The COTS program offset funding for both the Falcon 9 and Dragon development in support of cargo to the ISS, and if the entire system started flying successfully, the NASA team would gain more confidence in their capabilities, which could lead to trusting them enough to carry astronauts.

I saw the tide begin to turn in that direction when SpaceX started successfully delivering cargo to the Space Station in 2012. Having Dragon dock (or, in their case, berth) with the Station was a huge hurdle for NASA to overcome. If something had gone terribly wrong, not just the $150 billion Station was at risk; the astronauts on board would be put in danger, too. The Russians experienced a few hard dockings and near misses on earlier space stations—one that led to an evacuation of the module. I'd tried the Shuttle docking simulator myself and crashed every time.

SpaceX's second operational cargo mission launched smoothly into orbit but experienced a problem with its thrusters after separating from the Falcon 9 rocket. I was at the Cape for the launch and planned to meet up with Gwynne Shotwell for a bite afterward. She texted to let me know she would be delayed and invited me to their operations center, where they were working through the problem. After making sure the VIPs we'd hosted for the launch were safely back on their buses, I drove over to SpaceX to wait for Gwynne.

The difference between the SpaceX and NASA cultures was obvious as I entered the building. SpaceX's launch operation center looked like a double-wide trailer compared to NASA's massive, glassed-in launch control center, and it was now cramped with every console taken. Gwynne and I talked for a second, but I didn't want to bother her, so I headed to the back of the room to wait and watch. Joining me in staying out of the way was Gerst and the head of NASA's Space Station program, Mike Suffredini, known as "Suff."

Gerst and Suff were opposed to having the private sector take a leading role in human spaceflight when we initially proposed the idea. They were part of the leadership team that preferred government owned and operated systems like Constellation, SLS, and Orion. We'd battled over commercial partnerships for crew programs, but I knew it wasn't going to be successful unless they got on board. I had tried not to push them too far or too fast and kept the lines of communication open. It was understandable that they were more comfortable with traditional contractors, since that was how it had always been done. Between them, they were responsible for all of human spaceflight at NASA. They were in control.

Given this, I was surprised to see them watching from the back of the room instead of participating in the effort to fix the problem. Just one of the four thrusters was working and four were needed before Dragon could be cleared to approach the Station. There were several operational constraints to finding a solution, and time was running out. We were crowded together watching one console, and I overheard Gerst and Suff whispering to each other about possible ways to address the problem. I listened to them discuss potential fixes while watching the clock tick down until the opportunity to berth with the ISS would be lost, so I suggested maybe they should offer their ideas to someone at SpaceX.

Gerst calmly stated that SpaceX needed to figure this out for themselves. He and Suff seemed to be enjoying watching SpaceX work through various options while assessing how they were responding to the anomaly. I was sweating it out while my colleagues observed the action without visible stress. At one point I made Gerst promise me he'd intervene if it got too close, but he continued to give SpaceX the time.

SpaceX ultimately solved the problem without Gerst's and Suff's input and there were cheers all around. I relayed the story to Gwynne later that day, explaining why I thought it was an important sign that NASA leadership had rounded the corner.

What I'd witnessed felt like watching a grandparent and grandchild more than a parent and child. I like to describe it as if it were a fishing trip. A dad would try to help his kid bait the hook and learn to cast, but might take over the rod if the youngster caught something big and was struggling. Gerst and Suff let SpaceX select the best worm and put it on their own hook. They watched as the team cast their line in different places to figure out where the fish might be waiting. When the big fish started to bite, they let SpaceX reel it in on their own.

I'm not a grandparent but have been told how rewarding the role is by those with experience. They say it is all the good parts of parenting without any of the bad parts, since you have the maturity, trust, and patience to let your grandkids be who they are. I still get emotional thinking about the day I first saw the trust and patience NASA developed in SpaceX's abilities. The relationship had matured to a point where we trusted them with our most valued asset—the future of human spaceflight.

In 2014, before the final certification selection, rumors again circulated that Gerst, the chair of the source selection board, was considering selecting a single competitor and that it was Boeing. Although SpaceX had significantly outscored Boeing in most criteria, Gerst and others were concerned that SpaceX's low bid was unrealistic. If the rumors were true, he was either directed or had a change of heart. NASA awarded $4.2 billion to Boeing and $2.6 billion to SpaceX, as two fully funded Commercial Crew companies. Gerst later went to work for SpaceX, and Suff founded a company to build a commercial space station.

• • •

Working to get Commercial Crew established at NASA was a top priority, but there were certainly others. Nearly two years after President Obama announced that an asteroid was our next intended destination for astronauts, NASA had yet to come up with a program. Similar to the first budget process, Charlie appeared agnostic, which allowed NASA to be unresponsive to the President's direction. After being the Agency lead on Commercial Crew, I was aware my advocacy of another Presidential program without NASA buy-in would be intractable. My quest to get the Human Exploration Office that Gerst led to develop an asteroid mission was fruitless, until Charles Elachi came to see me with an idea. He pitched utilizing the technology demonstration missions we'd already begun, to rendezvous with an asteroid robotically and tug it into a position that could be reached by SLS and Orion.

Our original plan to have astronauts travel to a distant asteroid would have done more to drive advances in radiation protection and other human-sustainability technologies, but I couldn't make NASA come up with a mission and had grown weary from trying to push recalcitrant boulders. The mission Charles proposed had pluses and minuses, but the big plus was that I wouldn't be its only advocate. His enthusiasm was infectious, and I was thrilled to finally get traction on a program that aligned with the President's stated direction.

The mission would require a whole-of-NASA approach. The asteroid detection team was needed to develop more advanced ways to find and track one of the biggest threats to humanity. The technology team would have to (a) develop ways to keep them from crashing into Earth, (b) study the potential for utilizing them for future materials processing, and (c) test solar electric propulsion, rendezvous, capture, and tug technologies. The scientists would get an up-close look at a very large, pristine sample of one of the most important and mysterious heavenly bodies thought to have carried life throughout the galaxy. And SLS and Orion would finally have a destination.

One of the most elegant and appealing aspects of the mission was that NASA could obtain most of the technological and scientific benefits, even if SLS/Orion wasn't successful. On the flip side, if any of the technologies failed to reach, capture, or tug the asteroid into a position where it could be reached by Orion, the human spaceflight program would be no worse off than the current plan. As we began to socialize the idea within NASA, it started gaining support. Even Gerst and the new head of NASA's Science Directorate, John Grunsfeld, at least appeared to be onboard with the concept. I'd helped have Chris Scolese transferred to lead one of NASA's Centers in early 2012 and recommended one of our best Center directors, Robert Lightfoot as his replacement. Robert's support eliminated internal dissension. Having achieved unanimity, Charlie backed the plan enough to give me the go-ahead to try to sell it to the administration.

Elated at finally having a sanctioned project, I pulled together a team and created a strategy to pitch it at the White House. We set up a meeting to present the concept to Dr. Holdren, and when he became enthused, I was asked to brief the other senior staff in the EOP, while he ran it by the President. All parties were on board within a matter of weeks. I'd briefed the President's science advisor, not the President, but it was my mini version of President Kennedy accepting NASA's proposal to go to the Moon.

With the White House endorsement in hand, we presented it to the entire NASA leadership team. The combination of initial supporters and the substance of the concept earned their backing. Mike Coats had recently retired, and the enthusiastic support of his replacement, Dr. Ellen Ochoa, was especially appreciated.

Within the same time frame, OSTP initiated a government-wide program it called Grand Challenges. Like the X-Prize, it was structured to tap into the value of starting with a meaningful end-state goal that could attract the best minds in the world from the government, academia, and the private sector. Working with OMB, I learned there would be new money for Grand Challenges—meaning they would award additional funding outside the Agency's planned top line for accepted projects. The administration asked for proposals from all departments and agencies, and I pressed NASA to respond.

Charlie wasn't as enthused, but again let me ride point. The team I led proposed two concepts: one for studying Earth and one for studying asteroids. We presented them both to senior management. All agreed that funding the asteroid challenge aligned best with our immediate priorities, deferring the Earth challenge to a subsequent year. The title of our asteroid challenge briefing was: Be Smarter than the Dinosaurs.

NASA's proposed challenge was also quickly accepted by the White House. The Agency finally had a plan that could fulfill the President's direction from two years before. The first step would be to select a proper asteroid that could be retrieved and moved to an accessible orbit. The budget for asteroid and comet detection was $4 million when we started, and the Grand Challenge allowed us to build a more resilient and effective program that increased funding more than an order of magnitude to $140 million.

As we turned our attention to building public and political support for the mission, I proposed we select a name to help communicate more effectively. We held a meeting to discuss options between interested parties, and I recommended the name Artemis—the sister of Apollo. Orion was a Greek hunter and lover of Artemis, making it an appropriate name for the asteroid mission. As a demonstration of good faith, I suggested naming the SLS Zeus, the most powerful of all Greek gods. There seemed to be a general consensus around the names by those at the meeting and I prepared to take the recommendation to Charlie.

I'd written a paper in grad school about the social and economic implications of a lunar base, and had named the base Artemis. I was thrilled to finally have the sister of Apollo earn her place among the NASA stars. Unknown to me, in some versions of Greek mythology, Artemis mistakenly kills her lover Orion after being tricked into believing he is someone else. A rumor circulated that my motivation for recommending the name Artemis was a move against Orion and that naming SLS Zeus was a ploy to assure the rocket appeared last in alphabetical lists. These were not rational concerns. The fallacy was more likely circulated to keep the mission from gaining momentum and it worked. Even the hint of a lack of consensus kept the Administrator from making a decision, so the SLS was never given a proper name and the Asteroid Redirect Mission became known as ARM.

When the next administration named their proposed human spaceflight mission Artemis—and the Orion capsule was part of its mission—I didn't hear about any pushback from NASA's Greek mythology experts.

Nevertheless, developing and championing the asteroid mission was an extremely positive and collaborative experience. We had a trusting leadership team on an aligned goal that tied NASA's capabilities together to attempt a meaningful mission.

Similar to his hesitancy to support Commercial Crew in the beginning, the Administrator was either unwilling or unable to convey the rationale or purpose of the asteroid mission. Charlie's reticence gave an opening for people who had a self-interest in pursuing other ambitions to discredit the plan. The mission was estimated to cost an additional $3 billion, and that couldn't buy the support it needed on the Hill. Similar to lobbying on human spaceflight missions, contractors and researchers at universities advocate to their congressional delegations for directed funding. Landing on Mars was Charlie's focus and the big-ticket item he preferred to champion. His favorite tag line was that "we are closer to landing on Mars than ever before." The passage of time means this will remain perpetually true, with or without real progress.

Although the Asteroid Redirect Mission was developed to deliver numerous public benefits, it fell victim to the same forces that kept other innovative programs from proceeding. It didn't offer enough lucrative cost-plus contracts for industry and the Hill to support. While it served many different constituencies, it wasn't the top priority for any of them or for the NASA Administrator. The mission was designed to utilize SLS and Orion but had always been a forced fit, since most of the substantive benefits could be gained without sending astronauts.

The group of scientists studying asteroids and comets had learned to stick to their knitting and keep their heads down, not wanting to become a target for larger programs. Eventually the "small astronomical bodies" community supported the initiative, but it was never enough to overcome the more heavily funded lunar and planetary scientists who didn't want funding competition. Focusing on asteroid detection and deflection offered NASA the potential to align human spaceflight with more relevant public interests, but that wasn't the ambition of those with their hands on the switch.

10.

TURNING WRONGS INTO RIGHTS

AT AN ALL-HANDS MEETING WITH EMPLOYEES AT THE JOHNSON SPACE Center in the summer of 2010, Charlie Bolden compared the Constellation program to a stillborn baby calf extracted from a camel's womb by US Marines. Charlie said, "We've got some stillborn calves around, and we have got to figure out ways to help each other bring them back to life." The session was aired live on NASA TV and became a widely distributed meme. Charlie later confirmed that he was referencing past human spaceflight programs that had been canceled before they had a chance to succeed. I don't know why he chose camels or marines, and it is an atrocious metaphor in any circumstance, but it signified an important distinction between our perspectives. Charlie's goal was to resuscitate programs that had been canceled, and I was trying to understand and fix the systemic reason for their deficiency—the health of the mother camel.

Of the dozen human spaceflight programs NASA proposed since Apollo, only two had been completed. The Space Shuttle and International Space Station experienced years of delays, cost overruns, and tragic loss. Although both fell far short of their designated aspirations and programmatic expectations, they are viewed as successful because they were eventually completed.

Not only did their multibillion-dollar annual carrying costs make it nearly impossible to carve enough money out of the budget to develop something new, those banking the billions for operations had only disincentives to supporting potential replacement programs. The community complains that we haven't followed up Apollo with much progress in human spaceflight, but too many people have been invested in ignoring the true culprit.

Commercial Crew succeeded because it was a cart designed to follow the horse. Instead of designing the program around existing people and facilities, we hitched our wagon to the national benefits that would be derived from its success. We can now learn from the experience and adopt changes that lead to future healthy offspring. Incentivizing the private sector to lower space transportation costs was just the tip of the iceberg. Larger shifts are needed.

Fundamentally, we still have a system that creates programs to suit its own needs—doing the things the people who already work there want to do—instead of creating programs that could better address broad public purposes. Public polls on space issues consistently rank studying Earth science and detecting asteroids as the top two priorities for NASA, while sending a few humans to walk on the Moon and Mars rank at the bottom of the list.

Aligning NASA's goals to address today's challenges would expand interest and talent beyond its existing narrow constituency and encourage a more diverse workforce. Diversity builds stronger, more resilient and successful life forms in nature, just as it does in human collaboration.

As usual, Gene Roddenberry said it best: "We must learn not just to accept differences between ourselves and our ideas, but to enthusiastically welcome and enjoy them." An important justification for human spaceflight is its ability to inspire. Instead of trying to refuel Cold War hysteria for space stunts, we should be designing missions to inspire those NASA has yet to reach.

Gender and minority diversity in the astronaut corps wasn't a consideration for NASA in the early years—a fact often shrugged off as a consequence of the times. Reflection and context once again reveal something different. Civil rights, women's rights, and anti-war protestors marched arm-and-arm throughout the 1960s as the first seven classes of white male astronauts were selected one after another. A former NASA physician even ran a private selection process in the early 1960s to test female astronaut candidates. A group of women who became known as the Mercury 13 met the qualifications but were kept out of the NASA program. Women lobbied the White House and Congress to be able to compete, but NASA pulled out the big guns to defend its decision. John Glenn testified in 1962, "It is just a fact. The men go off and fight the wars and fly the airplanes and come back and help design and build and test them. The fact that women are not in this field is a fact of our social order." NASA and its heros didn't acknowledge it was a civilian agency.

In 1968, NASA was still holding Miss NASA beauty contests among its employees, with a lucky contestant chosen as Queen of Outer Space. One of my staff found a memo in the files from 1970 that read, "To: All Goddard Gals. Subject: Pantsuits.—On one side of the coin, I have to face you girls and your desire to be 'mod,' and on the other, the male population who would only vote for minis." The memo ultimately deems the new wardrobe item acceptable, but only "if you feel that a pants suit would not be offensive to your boss and would not embarrass him." In conclusion, the gals needed to bear in mind, "if someone forgets to treat you like a lady—it was you who elected to wear the pants." I can only imagine what women like Nancy Grace Roman, who was NASA's chief of astronomy at the time, thought of the memo.

Girls had to wear dresses when I first started in my public elementary school in 1966, except for gym days, when we were allowed to wear shorts underneath. I looked forward to gym days until the policy changed when I was in third grade—before the women who worked at the most futuristic public agency in the world were allowed to wear pants.

When *Hidden Figures* was released in 2016, most of us were unaware of the many Black women who worked at NASA in the 1960s. The film was based on the recently published book by Margot Lee Shetterly, who had grown up in the Virginia tidewater area and had known the families of the women chronicled in her book and the movie. Theatergoers in the Virginia suburb where I saw the movie laughed while watching our heroine run between buildings in order to use the restroom deemed suitable for colored girls, but nothing about that reality was funny or nostalgic.

The so-called computers at Langley weren't the only unsung women contributing to NASA's early success. Similar cadres of professional women were employed across the Agency in the 1960s and 1970s. Their stories had also largely gone untold until Nathalia Holt wrote *Rise of the Rocket Girls*—also published in 2016—about the female rocket scientists who worked at the Jet Propulsion Laboratory in the same era.

Public views on human spaceflight have been shaped predominately by men who had a natural instinct and affinity for NASA's early programs. There have been hundreds of books written about the history of spaceflight from their perspective. These historian's similar identities have had an outsized impact on what we think we know about the past. It shouldn't have been surprising that Lillian Cunningham's *Moonrise* podcast offered a different perspective on the forces behind NASA's formation.

Many relevant messages are finally being conveyed through the stories of NASA's women. The exchange between Dorothy Vaughan and Mrs. Mitchell, the white supervisor of the computers from *Hidden Figures*, offers a universal lesson. Attempting to defend her disrespectful management style, Mrs. Mitchell tells Dorothy, "Despite what you think, I have nothing against y'all." It's Dorothy's response that resonates with me most: "I know you probably believe that." Unintentional gender and racial bias are difficult for all of us to acknowledge and overcome.

Columbia University's 2003 "Howard vs. Heidi" experiment confirms this reality. The business school distributed two sets of background information to students to evaluate as potential employees. The two portfolios were identical, other than their names. Half the class evaluated Howard's qualifications and the other half evaluated Heidi's. The results are often reported as shocking, but to me, they are depressingly all too familiar.

Howard is the overwhelming favorite and is most often selected for hire over Heidi, even though their backgrounds and resumes are identical. Heidi is seen as selfish and conceited; Howard is seen as confident and powerful. Both are described by the students as assertive, but Heidi is rejected for the trait, while Howard is admired. The negative correlation between power and success for women is the exact opposite for men. The double standard I've lived—especially late in my career—is borne out by the study.

Like many professional women, I have struggled to find an acceptable approach to being an effective female leader. Upholding gender-based expectations requires accepting ideas when you disagree, being talked over in meetings, not objecting when men are given credit for your ideas, and taking on emotional labor at home and the office lest risking being perceived as selfish and thoughtless. The inequities most of us experienced back in the day were macro and micro. It is hardly worth recalling the number of times I was asked by men to do secretarial tasks or get them coffee, long after my administrative positions were behind me.

The space field was so overwhelmingly a white man's world that consideration wasn't given to how anything said or done might be offensive to women or minorities. The original *Star Trek* series is heralded for its diverse cast, yet exemplifies how women were viewed as sex objects, even projecting three-hundred years into the future. Gene Roddenberry proposed a female captain in the show's pilot, but the network would only pick up the series with a man in charge of the *Enterprise*. It wasn't until the third season—centuries later in the series storyline—that a woman was allowed to command the ship.

Women were portrayed as subjects who were there for the pleasure of the men. Gene shared with me that the execs pushed for females—human or alien—to show more skin every season. The only woman regularly allowed on the bridge—fan favorite Nichelle Nichols—was depicted as a one-dimensional telephone operator with few lines and a short skirt.

A recent documentary about Nichelle's life, called *Woman in Motion*, highlights that the NASA official who first reached out to her did so to find out for himself if her legs were as stunning in person as they were on TV. This introduction led to Nichelle's involvement in helping the Agency recruit women and minorities to the astronaut corps in the 1970s, which helped diversify the first class of Shuttle astronauts. The anecdote is portrayed in the 2021 film as a humorous happenstance, without acknowledging the sad irony of a role model for future female space professionals being initially sought out by a man for her looks. It serves as a poignant reminder that the times haven't changed enough.

Nichelle is a long-time board member of NSS, and I treasure our friendship. We were the only two women at our early board meetings. I could not have asked for a better role model or mentor. *Woman in Motion* is largely based on interviews with people who were inspired by

Nichelle, and I am one of only nine women out of the thirty-five people interviewed for the documentary. To quote Nichelle, "Where are my people?" NASA is still in need of more women in motion.

Being objectified was a part of being a woman working in aerospace when I was in my twenties and thirties, and I learned which men to avoid in the community. Many of us encountered unwanted sexual advances and harassing behavior without showing offense. On one of my birthdays in my thirties, my NASA supervisor in the policy office—in front of several other colleagues—told me to come into his office so I could get my birthday spanking. One of the more well-known harassers, a professor whose unwelcome sexual advances toward his younger female colleagues and students are legendary, was agitated at being rebuffed by a student. He told a prospective employer who was interviewing her for a job, "If you expect to sleep with her, don't hire her." The employer hired her anyway and later relayed the comment to her.

Being propositioned and groped by male colleagues was not uncommon and almost always came from older, more senior men in the industry. I was in Moscow during my first tour of duty in NASA (married and in my thirties), when a senior aerospace contractor who had been over-served pushed his way into my hotel room, shoving me onto the bed. I was able to get out from under him and run into the hall, finding a colleague to intervene. I never reported the incident to NASA or to his employer. Embarrassed and assuming it would be my own career that suffered, I—like so many others—swept such occurrences under the rug. I'm ashamed for many reasons, but mostly because the behavior likely continued.

Evolving from old ways of building rockets has come easier than embracing diversity, equity, and inclusion. Seventy-three white men were welcomed into the astronaut corps before a single woman or minority was chosen. Even as of this writing, no woman has led NASA in its sixty-five-year history. Selecting leadership and astronauts that are not representative of the majority of the population disenfranchises the public and perpetuates a vicious cycle. Change requires intention.

NASA receives many thousands of applications for each astronaut class. The demanding requirements and high standards aren't met by all applicants, but after down-selecting to the top 10 percent, nearly a thousand extremely qualified individuals remain. Only the top 10 percent of those are selected for in-person interviews, and only 10 percent of those become ASCANS—the astronauts' affectionate name for new astronaut candidates. Most astronauts go through the application process a few times before selection.

If NASA wants its rationale to be more than rhetoric, we need to be intentional in our selection criteria, especially since the corps has had such a preponderance of white males. In my view, unless and until our astronaut corps and crews represent society, we are falling short.

It is probable that more than .01 percent of candidates could become outstanding astronauts. Selecting between a helicopter pilot with a 3.8 grade point average in astrophysics from West Point and a fighter pilot with a 3.9 grade point average in chemical engineering from the Air Force Academy is, in many ways, subjective. If one of them is a white man and the other is a Black woman, selecting the latter candidate would align with a criterion to inspire segments of the public who have not historically been equally represented in the astronaut corps.

NASA announced the selection of nine astronauts less than a month before Charlie and I were confirmed. It was the first new astronaut class in five years. Group 20 included six white men, two white women, and one Black woman. I thought it was odd not to have received the information in advance, since we were taking home large binders filled with much less important data each night, and I was disappointed in the lack of diversity of the class. I brought it up with Charlie, but he didn't appear to share either of my concerns and suggested I talk with Mike Coats—the man who made the selections.

Approaching Mike on the topic seemed to make him uncomfortable. Instead of addressing his concerns with me directly, he sent a team from JSC to meet with me. Their briefing focused on data that showed less than 30 percent of the applicants had been women and less than 10 percent had been Black, noting that the race-based number was representative of the percentage of Black Americans in the population. I shared my view that the final selections' being representative of the candidate pool shouldn't be the benchmark, since we wanted to reach a segment of the public that had already been disadvantaged and therefore may not have applied. I thought we had a good discussion and appreciated them taking the time to share their data.

New astronaut classes and crew selections have consistently favored white males. Of the thirty-two astronauts selected, trained, and picked to fly on the six Shuttle missions under our watch, twenty-eight were men and thirty were white. An additional thirty astronauts were assigned or flew on Soyuz during President Obama's (and Charlie's) tenure. Twenty-five were men and five were women—only one a person of color. Combined, less than 15 percent of the astronauts who flew during the Obama-Bolden administration were women and less than 5 percent were people of color.

A month after our Senate confirmation, JSC announced the assignment of an all-white male crew to the next mission, STS-134, along with disclosing that the sole woman on the STS-132 mission was being replaced by a man, which made that one another crew of six white males. I was beyond frustrated. Dan Goldin had overseen sixty-five successful Space Shuttle missions during his tenure and only five had consisted of all-white male crews. Ten years later, the first Black Administrator and second female Deputy Administrator were overseeing a regression in astronaut diversity.

I put a marker down that four white males should not be assigned to the final Shuttle flight, STS-135. My plea was based on a view that one of the Shuttle era's contributions to human spaceflight was our transition out of the male military test pilot bastion of the early program. I was relieved when I learned that Sandy Magnus had been assigned to the crew, but disappointed that we missed our chance to have our final Shuttle mission be the first to represent gender equity.

Our initial chance to make a selection of an astronaut class came in 2011, when we recruited Group 21. The same team from Houston came up to give me a briefing, and I could tell they were excited to be delivering the news in person that it would be the first astronaut class to represent women and men equally. I was equally enthusiastic about the outcome and by the qualifications of the ten extraordinary individuals they selected. The most recent class, named in 2021, is 40 percent female.

I discovered while researching this book why my assertions were ignored. Mike Coats put it this way in a 2014 interview: "The Center director has final approval for crew assignments. Obviously, a NASA Administrator could override him if he wanted to, but it never happened with either Mike or Charlie. Got a bunch of questions when Lori Garver came in as Deputy Administrator. She questioned every crew assignment. How come we didn't have more minorities and more women? I'd usually let Charlie deal with that out there. Charlie frequently said, 'Don't worry about it.'"

Mike's words explained why my efforts became another Sisyphean battle. This was a confounding and disheartening discovery.

For many of the cup boys, gender and racial equity are not substantive issues; they are "political" issues. Mike detailed his view, saying:

> Within NASA, I think the problem that we had, is what I mentioned before, which is that this administration came in, and Lori Garver, and her focus was everything's political. What's good for the Democratic Party? Every decision. Does it help the union, or does it help the Democratic Party? Charlie was a very good buffer about that whenever Lori would question why there weren't more women and minorities on a crew. We'd explain why we assigned the crew. Sometimes there were more women than men. Sometimes there were more minorities than white people. Sometimes there weren't. If you want a quota on every crew, tell us, we'll certainly do it that way, but it's not the way to have the strongest crew. Charlie was very good about that. Her focus was what's politically correct.

Needless to say, there has never been a mission with more women than men or more minorities than white people.

A lack of diversity in the astronaut corps is something Charlie has often spoken about since he left NASA, but he was the singular person with the power to make meaningful change for nearly eight years. Instead of standing up for his conviction—or for his deputy—he chose not to rock the boat. Even when our goals were aligned, he couldn't give me his support.

Astronauts don't generally talk out of school, for fear of being grounded or ostracized, but their private stories about discrimination—especially in the early days—can be chilling. NASA's mishandling of the first class of women in space is known through humorous stories like men packing more bras and feminine products for a weeklong Shuttle mission than would have been needed for a month, but it hides a more malevolent reality.

Take NASA's relationship with the first American woman in space. Well beyond her initial fame as an astronaut, Sally Ride left an indelible mark on the Agency. She was the only person to have served on both Space Shuttle accident investigation boards—an experience that gave her deep insight into the workings of the bureaucracy and contractor relationships. Having fourteen colleagues and several close friends killed in the accidents, her depth of interest in what caused the catastrophes led to her profound and public frustration with the Agency's leadership.

Post-Challenger, Sally was tapped by NASA to lead an effort to lay out long-term goals for the Agency. The Ride Report, as it came to be known, formulated four scenarios: Mission to Planet Earth, Mission from Planet Earth, Outpost on the Moon, and Humans to Mars. The draft report recommended a focus on Mission to Planet Earth, but the Agency forced her to remove the prioritization before it was released.

Sally resigned within months of the report's publication in 1987.

Climate issues were just beginning to gain attention, but NASA wasn't about to publish information that suggested human spaceflight might not top its list of future priorities.

Sally spent most of her remaining years focused on outreach and education, with a goal to increase interest in science and space. Sally Ride Science was founded in 2001 with an early camp for girls that expanded to all genders over time.

Upon Sally's death in 2012, her partner Tam O'Shaughnessy revealed their twenty-seven-year relationship in her *New York Times* obituary. The news was shocking to many people in the space community, even many who knew Sally well. It was my privilege to work with Sally over the years and although we never discussed anything so personal, this piece of the puzzle fits.

Learning that her personal partnership with Tam began before her marriage to fellow-astronaut Steve Hawley ended led to speculation about her relationship with Steve and whether NASA had encouraged their nuptials. The wedding took place three months after her announced selection to be the first US female in space. It was sudden, private, and the only photo released of the ceremony shows the couple standing side by side in polo shirts and jeans. Sally's jeans were white. She and Steve divorced four years later, within months of her resigning from NASA.

Sally's family and close friends confirm her romantic relationships with women dated back to when she attended Stanford in the early 1970s, which has also raised questions about when this became known to NASA. In the official Sally Ride biography, author Lynn Sherr quotes Sally's husband of four years, Steve Hawley, as believing their marriage was sincere, but being at least somewhat unsurprised by the announcement in her obituary. It is possible her private life contributed to her reticence to serve as NASA Administrator in 1992 and 2008, but public views on LGBTQ+ individuals had dramatically evolved from 1978, when she was selected as an astronaut, and would not likely have been an issue for either the Clinton or Obama administrations.

At the time of her death, Amy Davidson Sorkin wrote in *The New Yorker*: "There is a valid historical question about what Ride's romantic life meant to NASA. Did you need to be married to a man to be the first woman in space? A number of press reports mention, by way of explanation for Ride's silence, that NASA would never have kept anyone openly gay in the space program."

Sadly, I do not disagree. Being a senior woman at NASA is already a tough road and can be made even more difficult by having additional differences with other members of the Agency. Being a scientist and astronaut helped Sally to fit in and do her work, and it is not at all surprising that she didn't want to go out of her way to highlight her less traditional private life. Sally was opinionated, direct, and outspoken in other matters. When people ask about my own personal mentors, she tops the list.

Sally was her own kind of space pirate. She didn't buy into the standard rhetoric about nostalgic human spaceflight or gender stereotypes and was deeply critical of NASA's decisions that led to the Shuttle accidents. Her career and support gave me headroom to be who I became as NASA's deputy. I couldn't have accomplished half of what I did without her blazing the trail, but she would be the first to admit that the brush was not yet fully cleared.

As president and COO of SpaceX, Gwynne Shotwell is an effective and revered female leader in the space community. Blue Origin lacks a similarly visible woman in its most senior ranks, but both companies have reputed "bro" cultures. Numerous charges of sexual harassment and discrimination have recently been made public by employees in both companies. Complaints and accusations of a toxic atmosphere that tolerates such behavior should not be ignored. It is time to end justifications for rooted misconduct as well as the field's predominance of people—including in its leadership—who look and think the same way. Progress toward diversity, equity, and inclusion has been much too slow.

Supporting women and minorities is becoming a tradition in aerospace, as more of us are determined to see future generations represented more equally in the field. Being a mentor to early career women and gender minorities is the most rewarding aspect of my later career. One of the women I mentored was Dawn Brooke Owens. We first got to know each other when she worked at the FAA on commercial space. Our bond grew stronger when she took a job in the White House under President Obama and was assigned the NASA account at the Office of Management and Budget.

Brooke was diagnosed with breast cancer on her thirtieth birthday and bravely survived for six years. No amount of time would have been enough for Brooke to accomplish all she aimed for, but it was tragic to lose her so young. Being among only a handful of women in the male-dominated aerospace industry, Brooke and I formed a close connection. We didn't want to just fit in; we wanted to bring our differences into the room. We talked about how to recruit more women into aviation and space over the years, and she and I were wired to do more than talk. The day after Brooke died, I drafted a stream-of-consciousness email with an idea to start an internship program for collegiate women interested in aerospace and forwarded it to a dozen colleagues. More positive developments followed than I ever could have imagined.

Many people responded to the message with suggestions and offers of assistance, but two of Brooke's best friends—Will Pomerantz and Cassie Lee—were all-in right out of the blocks. Six years into the program, more than two hundred women and gender minorities have become Brookies—the name they chose for themselves. We receive hundreds of applications annually for forty paid internships at host companies across the aerospace sector. Each Brookie is assigned a mentor and is supported by a growing cohort of peers and professionals. Everywhere I go, I meet students or recent grads who have applied, and even if they weren't accepted, they are quick to tell me how much it means to them that the program exists.

Tragically, the space community lost another young member of our family in 2017, Matthew Isakowitz. Matt was an engineer, entrepreneur, and extraordinary individual whose passion for commercial space exploration led him to work at the Commercial Spaceflight Federation. Matt was a confidant and key advisor to me during the most difficult early battles for the Commercial Crew program. Even in his twenties, his efforts were critical to our success. The Matthew Isakowitz Fellowship leveraged our templates and contacts to initiate a program focused on developing the next generation of commercial spaceflight leaders. Now in their fifth year, the 2022 class of Matt Isakowitz Fellows are part of a growing family that is bringing new energy and insight to the field.

Most recently, a third leg of our stool has been created, the Patti Grace Smith Fellowship. Patti was an aerospace trailblazer, a pillar of the Black community, mentor to Brooke, and a friend to me and many others. She was one of the students to integrate the public schools in the Jim Crow South, and she spent the latter part of her career as the Director of the Office of Commercial Space at the FAA. Patti died of pancreatic cancer at age 68, just a few weeks before Brooke's passing. Patti left an indelible mark on our industry, and we knew it was time to honor her as we had Brooke and Matthew.

Our first two programs were opening doors to more young people, women, and gender minorities. But even though we were selecting Black students at a rate higher than the current aerospace industry, we realized we weren't doing enough. Again, the aerospace community rallied, and the second class of Patti Grace Smith Fellows have already begun their journey.

These three distinct fellowships provide paid internships and mentorship to more than one hundred aerospace students each year, and two focus on serving underrepresented communities. This growing cadre of young talent entering our workforce is already making a positive difference in the sector, and more programs are being developed.

I'm proud of the accomplishments I've made in my career, and this book is dedicated to one of the most meaningful—driving reforms at NASA that are leading to more valuable and sustainable space activities. But there is no question in my mind that creating these Fellowships will lead to more progress than what any of us could ever accomplish on our own. The ripples from exposing a more diverse next generation of people to the field, just as the opportunities for more creativity and innovation in space are expanding, will undoubtedly leave the biggest mark. As this new workforce advances into leadership roles, I hope their ideas will be judged on merit equally.

• • •

The initiatives I pursued were not radical and did not pose a threat to NASA, human spaceflight, or the future of our children and grandchildren, as Gene Cernan and others charged. Similar plans had been proposed by the Administrator of NASA ten years before. Had they again been championed by a man, the criticisms would likely have been more muted. Several of the male Deputies who served before me were direct and outspoken and respected for it. I was the Agency's twelfth Deputy Administrator, and came to the position with over twenty years of experience in the aerospace field, but being a woman without an engineering background kept some people from engaging in rational and respectful policy debates. Many who disagreed with my views attacked me with vulgar, gendered language, degradation, and physical threats.

I've been called an ugly whore, a motherfucking bitch, and a cunt; told I need to get laid, and asked if I'm on my period or going through menopause. Bundled emails were sent to members and staff on the Hill and throughout the aerospace community from a group calling itself Change NASA Now, aimed at having me removed from my position:

- The problem with Lori Garver is that she is not qualified for the position that she is in. The fact that the recommendations she makes are given the weight that they are is going to be very destructive to the ability of the United States to get to LEO and beyond.
- She's a political appointee with a dearth of actual space experience. She should be fired for recommending nuclear-like damage to our NASA manned space program for the next two decades, allowing the Russians, Chinese and India to take over space pre-eminence. Yea, thanks a lot Obama. Hope and change, all right.
- Lori Garver proposed the most amateurish, incoherent plan and budget in NASA [and her] misguided plan and its attempted implementation, parts of which are very questionable legally, led to the lowest morale in the history of the Agency and massive, unnecessary layoffs.
- Lori Garver clearly understood Congress would never accept her harebrained plan so did not include Congress while developing it, nor her own boss, NASA Administrator Charles Bolden. Since Lori Garver's fingerprints were all over the plan, Congress was able to quickly determine who the chief culprit was. Now is time for Congress to act and demand the President get rid of Lori Garver as a sign of good faith to Congress, NASA, and the broader space community.

Disagreements over the President's proposed plan by those who stood to lose billions of dollars in contracts were expected and understandable. Gender-based attacks and lobbying campaigns that spread lies in support of removing me for proposing "legally questionable" and "harebrained" policies and programs designed to purposefully damage NASA were not.

NASA servers intercepted several death threats against me that were analyzed by the FBI. The envelope filled with white powder that was addressed to me at NASA Headquarters in August of 2012 was not the isolated incident I had hoped. A security detail was assigned to walk me to and from my car in the NASA garage on days where the threat level was heightened. There were times when I could have been more tactful, but the personal attacks and negative reaction to my ideas were undoubtedly heightened by my gender.

So why did I do it? It would have been much easier and more personally pleasant not to look under the hood. Instead of revealing what we found, I could have done what others have done: papered over problems and soft-pedaled the human spaceflight crisis to the President and in the transition team report so there would be no need for an external review. I could have been the kind of government leader who defended what we were already doing—the kind of girl people like—someone who goes along to get along.

For me, that was never a consideration. As Jessica Rabbit says, *I'm just not drawn that way*. At an ethereal level I saw NASA's work—my work—as having potential positive consequences for the future of humanity. My daily motivation was undoubtedly less admirable. I knew I was right and that it mattered, and I became invested in both points being proven. I tried to project an image of relishing the battle, but being maligned and condemned by the most revered and respected people in the space community was excruciating and left deep scars. Most people, especially women, like to be liked and I am not immune to this desire. I would have preferred not to have made foes in the process but, given the stakes, it seemed like a relatively small price to pay.

Making enemies is an inevitable outcome of leading transformative change, and *Moneyball* is just a current take on a deeply rooted theme. An earlier version of the sentiment, entitled "No Enemies," was printed in Upton Sinclair's anthology of the literature of social protest, written by Scottish poet Charles Mackay in 1846:

> You have no enemies, you say? Alas, my friend, the boast is poor. He who has mingled in the fray of duty that the brave endure, must have made foes. If you have none, small is the work that you have done. You've hit

no traitor on the hip. You've dashed no cup from perjured lip. You've never turned the wrong to right. You've been a coward in the fight.

The transformative agenda I was driving at NASA was in no way small work.

11.

UNLEASHING THE DRAGON

I FORMALLY BEGAN MY WORK FOR BARACK OBAMA IN THE SUMMER OF 2008 with a goal to shift the trajectory of space development in a direction that could sustainably advance society. After five years, I was proud of the strides we'd made on many of the priorities set for the Agency, but the boulders I was still trying to push up the hill had gotten heavier. I was increasingly being marginalized by the Administrator and therefore was having diminishing impact.

I'd been told the White House was going to select a new NASA Administrator in the second term, but as spring turned to summer in 2013 without any such movement, I returned a cold call I received from a headhunter looking for a "game changer" to fill a senior position at a Washington-based aerospace association. Five interviews later, I was offered the general manager position at the Air Line Pilots Association.

After talking with a few of my closest colleagues, I realized that the biggest boulders I'd been able to help set in motion were over the hump—they had reached apogee. We were far enough into the journey that the momentum in the right direction would continue. The constant struggle had left me feeling like the *The Old Man and the Sea*; I had gone out too far, and I didn't want to lose myself or what we were working to accomplish. I drafted my letter of resignation to the President and accepted the job offer.

The reaction to the announcement of my departure was as expected. I was exalted by some and vilified by others. I was asked by reporters in an exit interview if the SLS would meet its then-2017 launch date, and I acknowledged it was likely to slip a year or two. Charlie had NASA put out a statement that said I was mistaken and that all was on track for its first test launch by the end of 2017. Boeing's SLS program manager, Virginia Barnes, confirmed NASA's view, saying, "I have not heard even rumors of slips on this SLS rocket. In fact, my schedule looks five months ahead of schedule. That's across the board." It's now planned for 2022.

I kept my eye on the Commercial Crew program after leaving NASA, following their test schedules, and cheering on both teams. Most observers assumed Boeing would launch first, and that is the way the flights appeared on the manifest initially. Boeing had received nearly twice the NASA funding, but late in 2019, the company experienced a major setback on what was to be their last uncrewed flight test. A software problem precluded docking with the Station and their capsule, named Starliner, has remained grounded ever since. Boeing thought they were ready to test again (at their own expense) in August 2021 but had to stand down after stuck propellant valves were discovered once the vehicle was on the launchpad. Boeing now hopes to run the test flight by mid-2022.

NASA leadership acknowledged that their greater familiarity with Boeing than SpaceX led to less oversight for the Starliner spacecraft. The manager of NASA's commercial program, Steve Stich, admitted in 2020 that the software errors that surfaced on Boeing's first test flight of the Starliner were a result of too little scrutiny based on the trust they'd developed over decades of working with them.

Boeing's misstep opened the door for SpaceX. Their final uncrewed test flight took place in January 2020—the in-flight abort test. In the event of an emergency, the Dragon capsule would need to demonstrate it could eject from the rocket safely, even if the rocket exploded at the peak moment of pressure during launch. The company passed the test with flying colors, paving the way for the full test launch, known as the Demo-2 Mission, which would carry astronauts to the Station for the first time.

SpaceX received more scrutiny and less money but still made it to the launchpad first. The mission was what many of the space pirates and I had been paving the way for over decades and what Elon and his team had been working toward since the founding of SpaceX in 2002. The launch was scheduled for May 27, 2020.

SpaceX and NASA had announced years before that astronauts Doug Hurley and Bob Behnken, both test pilots, were assigned to the mission. The astronaut duo harkened to NASA's fly boy past. They had been picked specifically for their extensive military training, as well as their deep connection to each other. Doug had been the pilot on the last Space Shuttle mission and the symbolism of his commanding the first return to flight was poignant. If their Dragon launch and docking was successful, they planned to remain on the International Space Station for a thirty-to-sixty-day stay, returning into the ocean in the same capsule on parachutes.

I thought nothing could keep me from going to the first Commercial Crew launch, but in May, with COVID raging, many of us reluctantly decided not to make the trip. Several of us from the core NASA and White House team who had charted the course on the policy side planned a Zoom call to watch the launch together remotely, promising to meet in person at the Cape for the first post-COVID launch.

NASA's coverage of preparations in the suit-up room showed the astronauts in new SpaceX Jetsons-like flight suits, talking with Elon Musk and Jim Bridenstine, who had been confirmed as President Trump's NASA Administrator just over two years earlier. I had attended Shuttle launches in person so regularly that it was enjoyable to actually watch what was happening on NASA TV. When the astronauts headed outside for the drive to the launch tower, they waved to their families as usual, but instead of getting into the classic NASA Astrovan, they rode in their own white Tesla Model X's. If anyone had missed the transition, the bit with the Teslas hammered the point home. This wasn't just the US Space Agency launching astronauts to orbit, but a private company. The electric cars were plastered with NASA logos—emphasizing the Agency's commitment and embrace of this new era of human spaceflight.

Ubiquitous NASA logos on the cars, rocket, and capsule were an apt symbol of the transformation at the Agency. SpaceX had wanted to put NASA's logo on the Falcon 9 rocket that launched their first commercial cargo mission to the ISS in 2012, but the space agency had initially declined their request. Gwynne Shotwell called to ask if there was anything I could do, and I had agreed to look into the matter. My first call was to the head of communications, hoping David Weaver would tell me there had been some low-level decision that could be fixed. Unfortunately, that was not the message. He said the direction had come from Gerst: no NASA logo of any kind was to be on the rocket or capsule. I talked to Gerst himself, who explained it wasn't our rocket, and the lawyers made the decision. I knew it wasn't a legal issue, but Charlie didn't want to press, so I had to call Gwynne back to let her know that I couldn't get it fixed.

Eight years later, the SpaceX rocket and capsule looked like a NASCAR vehicle covered in NASA logos. I even heard that Gwynne had to make a different call this time; she needed help getting NASA to use a smaller logo on the vehicle because what the Agency chose was so large that SpaceX was concerned the darker color could cause the spacecraft to overheat during reentry and fry some of the electronics. As I knew was the case in 2012, there was no legal reason NASA couldn't put their logo on a SpaceX vehicle. It was one of hundreds of "calls" made by bureaucrats with an axe to grind that the Administrator hadn't been willing to set aside. Administrator Bridenstine had no such compunction.

• • •

The nation was dealing with more than COVID in the spring of 2020. On May 25, just two days before the planned SpaceX launch, police in Minneapolis, Minnesota, murdered George Floyd after suspecting he used a counterfeit twenty-dollar bill at a convenience store. The video of a police officer jamming his knee in Mr. Floyd's neck and blocking his ability to breath went viral over the coming days. It was a brutal killing and it was one of a long history of unprovoked police violence against Black people. Many of us had had enough and took to the streets in protest. The Black Lives Matter movement gripped the nation and the nation's capital city. I joined thousands of people at rallies and marches, wearing masks and carrying signs demanding positive change.

The confluence of racially charged protests with the first Dragon launch inevitably led to comparisons to the 1960s. Jim Bridenstine was asked at the pre-launch press conference what the mission offered to a divided nation. He responded that he "hoped it could bridge a divide, but that if people thought this was going to end their problems, their expectations were too high."

The Administrator's answer acknowledged another aspect of NASA's mythology. The Apollo 8 Mission is remembered as providing a beacon of hope to a struggling nation, but while the first view of Earth from the Moon and spiritual message from the astronauts briefly lifted our gaze in 1968, it didn't end the war in Vietnam, nor did it bring an end to systemic racism and poverty. The pronounced cultural and racial divide of the times was memorialized in Bluesologist Gil Scott-Heron's poem "Whitey on the Moon" in 1970, which compared the experience of the Black community—a rat biting his sister and not being able to pay her doctor bills—with the experience of white astronauts going to the Moon. The poem ends with the lines, "Y'know I jus' 'bout had my fill of Whitey on the moon—I think I'll sen' these doctor bills, Airmail special, to Whitey on the moon."

Much of the mainstream media covered the SpaceX launch on a split-screen with news of the demonstrations in reaction to George Floyd's killing two days earlier. To me, it felt like the nation was coming apart at the seams. It was hard to focus on anything other than the hatred and violence that continued to be displayed by people in our country charged with protecting and defending our values. I couldn't help thinking it was unfortunate that the first mission of what was billed as *the next space age* launched two white men, making the program appear out of touch with current society.

The exploits of two or three people in space, no matter their race or gender, can't relieve the pain of people struggling to afford food, housing, healthcare, and basic rights; if anything, it shines a spotlight on a growing divide in society. Meaningful social change requires establishing permanent two-way bridges to heal our deep divisions. Lifting up those who have been disadvantaged requires intentionality. Diversifying astronaut crews provides role models and can be a ray of hope for people who rarely see themselves in such positions, but it is only a piece of the substantive reforms NASA could undertake to support a more just and inclusive society.

When Bob and Doug rolled out in their tricked-out Teslas headed for the launchpads with a NASA helicopter thundering overhead, I knew the visual takeaway didn't transmit the public value inherent in what was being achieved. Conveying the potential meaningful benefits to society that could be realized by reducing the cost of space transportation would have to wait for another day.

As I was watching the TV coverage, I saw one of NASA's former astronauts who wasn't going to space that day but had worked at SpaceX. Garrett Reisman was dressed in his unmistakable blue NASA astronaut jacket, standing by the roadway as Bob and Doug zipped by under battery power while jamming to their preselected playlists. Garrett cheered them on as one of the people who had helped build NASA's trust in SpaceX and embrace the company as a partner. While there were still cultural differences, what was about to happen represented hundreds of people like Garrett, working in a successful partnership for over a decade.

The astronauts arrived at the former Apollo and Shuttle launchpad and took the elevator to the top of the newly renovated, black-and-white launch tower, ready to board the Dragon. NASA announced the capsule was given the name Endeavor, the name of the Shuttle both Bob and Doug had previously taken to space. As the pair was strapped in by SpaceX employees, I received a text alerting me that a Brooke Owens Fellowship alumna was a member of the close-out crew. As I watched Maddie Kothe—number nineteen—help the astronauts strap in, I felt an even deeper connection to the mission.

One of SpaceX's lead engineers on Crew Dragon had sent me an email a few years before with the subject: "I need help recruiting awesome women." I'd recommended Maddie apply, even though she was weeks away from starting a master's program in engineering at Stanford. When she got the SpaceX offer, she called to ask for my advice. After talking about it awhile, she realized her goal in getting a master's was to be eligible for the type of job SpaceX was offering. Two years later, she was buttoning up the astronauts on the first commercial human spaceflight in history. It didn't look like she had any regrets.

The public lined the Florida highways and beaches in the hope of seeing history in the making, while the media and VIPs who made the trip to the Sunshine State wore masks. Even the press site at Kennedy, normally a hive of activity, was quiet. The weather was a coin flip when Bob and Doug arrived at the launchpad that afternoon. As the countdown ticked lower, the storm clouds arrived and stopped the clock at 16 minutes and 53 seconds. Liftoff was reset for three days later.

The weather looked like it might also be a problem on May 30, but SpaceX and NASA gave the go-ahead and preparations began again. We were casually reminiscing on the Zoom call, assuming there would be another delay, when we realized the countdown was proceeding. Suddenly, as if it was a movie, the storm front moved on and Mission Control started polling the team—"Go" "Go" "Go," one after another.

With the walkway pulled back, you could see the spacecraft balanced on top of the rocket. Bob and Doug seemed as if they were teetering on a thin knife, ready to split the sky. The rocket began venting as the cryogenically cooled fuel was pumped into its core. I was more nervous than I wanted to admit. I'd thought about this exact day for twelve years, knowing if something went wrong, I would blame myself. I hadn't touched a single piece of hardware, but having helped chart the course that led here, I felt somewhat responsible.

When each of Falcon's nine engines fired up, snapping into life in a bright flash at the base of the launchpad, the rocket began to move, and the power cords fell away. As the rocket rose—faster and faster—the engines roared. Media coverage continued split-screen coverage of Black Lives Matter demonstrations next to the Dragon capsule carrying Bob and Doug into the upper reaches of the atmosphere and disappearing from sight in about a minute. I couldn't distinguish which of the two events were more responsible for feeding my tears.

Bob and Doug continued on, powered by Falcon 9's upper stage. Inside the cabin, where the pair were strapped in, a sequined stuffed purple dinosaur floated into view. They had made it to space! The dinosaur was the microgravity indicator that Bob's and Doug's children had chosen to signify when they had reached the point of weightlessness. A euphoria of cheering echoed through the webcast: SpaceX employees celebrating loudly back at their headquarters in Los Angeles. The company had achieved something that many thought no private entity ever could and had earned its right to celebrate.

Recognizing that any rocket launch with people is only as good as its landing, I held off on making a full victory lap. Others were less circumspect. President Trump, Vice President Pence, and NASA Administrator Bridenstine understandably received and acknowledged credit on behalf of the government for the successful SpaceX Demo Mission. In media interviews and on Twitter, I joined in thanking them for responsibly carrying out the program. In typical Trump fashion, he overstated his involvement, but he could have canceled the program, and I am grateful that didn't happen.

Bridenstine tweeted, "Under President Trump's leadership, we are once again launching American astronauts on American rockets from American soil." Some complained about the presidential nod, but Bridenstine reported to the President, and it was a true statement. A comparison with Nixon taking credit for landing the first people on the Moon was self-evident to most people.

Bridenstine further stated, "This is a program that demonstrates the success when you have continuity of purpose going from one administration to the next," saying that the Commercial Crew program had built on the success of the commercial cargo resupply program started by former President George W. Bush nearly fifteen years ago. He also praised his predecessor: "Charlie Bolden did absolutely magnificent work as the NASA Administrator," he said, including selling the Commercial Crew program when it "didn't have a lot of support in Congress. Charlie Bolden did just yeoman's work in order to get this program off the ground, get it going."

I admire much of what Administrator Bridenstine tried to accomplish during his NASA tenure, and credit his remarks for showing bipartisanship. Perhaps he wasn't aware that the "continuity of purpose" began long before the George W. Bush administration, but Jim was well aware that it was me, rather than Charlie, who'd actually supported the program in the early days. Rewriting history to proclaim male heroes came more naturally, so NASA's patriarchal mythology was perpetuated. Being right had gotten us here, and that is what mattered.

Two months after the launch, SpaceX safely landed the Dragon spacecraft carrying Doug and Bob in the Gulf of Mexico, and I commenced my own personal rejoicing.

• • •

The next mission was the first operational Crew Dragon, called SpaceX Crew 1, and was scheduled to launch in November 2020. The complement of four astronauts, Mike Hopkins, Victor Glover, Soichi Noguchi, and Shannon Walker, named their vehicle Resilience. We'd all become more accustomed to managing COVID-era excursions outside the confines of our homes by November, and I couldn't stay away any longer.

My husband and I made the fifteen-hour road trip to Florida for the launch. Standing on the balcony at the viewing site where I'd gladhanded so many VIPs during Shuttle launches—now with Vice President Pence and Jim Bridenstine ten feet away—we watched Resilience rise into the dark sky. It was an honor to be invited as a guest, and I was overcome by a mixture of joy and relief as I listened to the roar of the rockets and then felt the rumble of sound waves as the four astronauts traveled higher and higher.

Congressional funding cuts and technical issues had led to delays, but the Commercial Crew program is the first of any NASA human spaceflight program since Apollo to be delivered within its planned budget and at a development cost that is an order of magnitude less than anything that has come before. Eight years and ten months after retiring the Space Shuttle, the furry mammals achieved their first success brilliantly. After years of building trust between them, NASA and SpaceX are working together as a team and the tsunami has begun to retreat.

SpaceX's success has started to blur the lines between former factions in the community. Even the US military acknowledges how much the reduction in cost and responsiveness of space assets have enhanced our national security. In his confirmation hearing, Secretary of Defense Lloyd Austin testified that the innovations of space entrepreneurs as a means of strengthening the military's hand was a uniquely American way of sharpening the military's edge. Other senior Defense Department officials have noted that the objective to rely less on federal teams than the tech entrepreneurs who were fast transforming the civilian world has paid off.

The New York Times reported in early 2021 that the NASA approach of funding numerous companies competitively instead of the usual method that dictated terms to contractors provides the United States its best strategic military advantage in space. The article read: "For Mr. Obama, innovative leaps were to do for American space forces what Steve Jobs did for terrestrial gadgets, running circles around the calcified ministries of authoritarian states." The paper credited NASA's relatively small investments in SpaceX, Blue Origin, and other entrepreneurial companies, as leading to a new unconventional edge in US national security. And the *Times* noted that the reusability of launch vehicles and reduced cost and size of satellites was allowing military planners to make anti-satellite targeting vastly more difficult—in some cases impossible—for an adversary.

Numerous friends and former colleagues forwarded the *Times* piece to me the morning it was published, knowing I would find it particularly rewarding. The denunciations from elected leaders charging that what we had proposed would undermine national security was never about a true concern for our collective future, it was about trying to preserve their own.

• • •

The privately funded space race in suborbital space tourism attracted the world's attention in the summer of 2021, delivering on its decades-long promise. Virgin Galactic and Blue Origin successfully launched and returned their founders to the edge of space and are now beginning to carry paying passengers as tourists.

Blue Origin was poised to win this particular heat when they announced that their founder Jeff Bezos and his brother would join the winner of an auction on their first crewed flight, scheduled for July 20, the 52nd anniversary of the first Moon landing. Rounding out the flight crew was one of the original Mercury 13 women who had been waiting for her opportunity since the 1960s, Wally Funk, and the son of a billionaire from the Netherlands who had purchased the ticket for an undisclosed price.

Virgin Galactic was still in their testing phase, but announced two weeks after Blue Origin's announcement that founder Richard Branson would be going on their next flight. It was scheduled to take place the week before Blue Origin's. Branson was joined by three members of the Virgin Galactic team in addition to the two pilots required to fly SpaceShipTwo. One of the staff members selected was a friend and former colleague who runs their DC office, so I was included on her guest list and made the trip out to Spaceport America in New Mexico for the launch.

Blue Origin was understandably miffed at being beat off the line after announcing their date first. They publicized a comparison between their vehicle, the New Shepard, and Virgin Galactic's SpaceShipTwo. They pointed out that the New Shepard takes off like a rocket, goes higher, has an escape system and larger windows. Blue Origin's message came off as mean-spirited, even to supporters. Memes were immediately circulated showing additional comparisons, such as which rocket looked more like a penis. Bezos tweeted a "best wishes" message to Branson on his personal Twitter account the day before the flight as well as congratulations when the flight was successful.

Elon Musk uncharacteristically stayed out of the fray on social media, other than announcing he'd be attending the Virgin Galactic launch. Elon arrived at the VIP site with his one-year-old, X Æ A-12, and his diaper bag a bit disheveled shortly before the flight. A tweet from Branson a few hours earlier showed a photo of him arm in arm with a barefoot Musk with the message, "Big day ahead. Great to start the morning with a friend. Feeling good, feeling excited, feeling ready." I later heard that Elon had turned up at the house where Richard stays when he's in New Mexico the night before and had slept on his couch.

Branson continued to say the timing of his flight just beating out Bezos was a coincidence. It was hard to believe, but if the preparations were really put together in such a short time, you could have fooled me. The event was staged to perfection. The Spaceport itself is an architectural marvel and served as the perfect backdrop to the show. Taking off and landing from a runway allows guests to be closer to the scene than they are allowed for vertical rockets.

The guests in attendance watched as the massive White Knight Two airplane took off with the SpaceShipTwo rocket strapped under its wings. After climbing to 40,000 feet, the rocket was released, its engine ignited and about a minute later, the newly minted astronauts gazed out their windows at the curvature of Earth while floating inside weightlessly. I was enraptured watching my friend Sirisha Bandla start testing procedures on the science experiment she flew on the big screen before she passed through the cabin doing somersaults. We all started breathing again when the rocket touched down about five minutes later. SpaceShipTwo was towed to the stage while the musician Khalid performed a song he'd written for the occasion called "New Normal."

Ten days later, it was Blue Origin's turn to light the candle. Getting to Van Horn, Texas, isn't any easier than getting to Truth or Consequences, New Mexico (home of Spaceport America), but the same throng of media, along with VIPs in Jeff Bezos's circle, gathered on the morning of July 20. Jeff isn't as much of a showman, so the event was less scripted than Branson's. Spectators had to be five miles away, but large screens provided real-time video. The excitement among the crowd was palpable as Jeff and his crew boarded the spacecraft.

The first stage of the vertical rocket jumped off the pad and burned for just over two minutes before throwing the New Shepard capsule into space. After their few minutes of weightlessness, the four members of the crew strapped in and landed on parachutes not long after or far away from where the first stage rocket had returned on a tail of fire.

Blue Origin flew a second tourist flight in October of 2021 and set a record for carrying the world's oldest space traveler, ninety-year-old William Shatner, the actor best known as Captain James Kirk in *Star Trek*, and three other passengers. Jeff Bezos welcomed the crew home immediately after their return from space and was told by Shatner, "What you have given me is the most profound experience I can imagine." The iconic actor was clearly moved. "This air which is keeping us alive is thinner than your skin," he said, adding, "It would be so important for everybody to have that experience, through one means or another." The suborbital tourist flights were just the warm-up act.

In the fall of 2021, SpaceX flew the first entirely commercial orbital mission without NASA involvement. Four new astronauts were minted after the three-day trip, all paid for by Jared Isaacman, a thirty-eight-year-old billionaire entrepreneur. Isaacman offered the remaining seats to individuals selected independently, with all proceeds donated to charity. The mission, called Inspiration4, included a lottery winner, a pediatric cancer survivor with a prosthesis, and an educational artist entrepreneur. Jared served as commander and donated $100 million to St. Jude Children's Hospital, an amount that was more than matched by others. It was the first space mission consisting of only non-government employees, and the first to fly an equal number of men and women. The Netflix mini series that coverd the flight was reminiscent of what the Discovery Channel had planned for Astromom twenty years earlier.

NASA is also finally beginning to facilitate tourist flights to the International Space Station. Former NASA employees started a company called Axiom Space, which serves as an intermediary with the government, and retired astronauts are being paid as guides for Dragon flights to the ISS. Two excursions with three tourists each are already on the books for 2022 at the advertised price of $55 million per seat.

• • •

Denigrating "space tourism" as a playground for the rich may seem like a fair criticism, but it deserves a deeper analysis. Tourism is nearly a two-trillion-dollar industry and returns huge economic benefits to nations that lure the largest shares of the market. It will likely take many years for space tourism to become a significant portion of that market, but the opportunity to have the United States lead the business shouldn't be dismissed. Suborbital spacecraft have the potential to carry cargo or passengers from one point to another anywhere on planet Earth in ninety minutes. That market may hold the most promise.

The emerging tourist market, combined with increased troubles on the now-aging ISS, and a need to reduce its approximately $3 billion annual operational costs, have hastened NASA's interest in transitioning to privately owned and operated Earth-orbiting laboratories.

NASA announced the Commercial LEO Destinations (CLD) program to co-fund the development of commercial space stations in 2021. Modeled after the Commercial Crew and cargo programs, NASA is offering Space Act Agreement partnerships, to be followed by service contracts by the end of the decade. Twelve bidders responded to its initial solicitation, increasing the likelihood that there will be more places people—whether they be tourists, scientists, or astronauts—can visit in the not too distant future.

• • •

There is much speculation over what the ultimate outcome of the Elon Musk–Jeff Bezos rivalry will be, and although it is too soon to tell, their trading places as the first and second richest individuals on the planet in 2020 and 2021 makes comparisons inevitable. Great rivalries throughout history are known to advance the state of the art beyond what one individual could achieve on their own. The "power of two" theory includes advances made by Da Vinci and Michelangelo, Edison and Tesla, the Wright Brothers and Curtis, Gates and Jobs. Having Musk and Bezos investing in accelerating sustainable space development at the same time has already changed the game.

Jeff Bezos's long-term vision for space development includes moving environmentally destructive industries off the planet to assure the survival of life on Earth, while Elon Musk's focus is on sustaining humanity on Mars. Even if these visions aren't realized for generations, their privately funded efforts are dramatically reducing the cost and simultaneously increasing the capabilities of satellites and space transportation, providing billions of dollars of economic benefit to the United States. Hundreds of commercial space companies are now advancing the state of technology in ways inconceivable just ten years ago.

As of this writing, Elon clocks in as the world's richest person, with a net worth of $336 billion, while Jeff has fallen to number two with a net worth of $196 billion following his recent divorce and partial divestiture. Amazon is bigger and its outsized impact on local businesses and the environment make it more controversial than Tesla, but the billionaire backlash has hit them roughly equally. Elon has a larger cult-like following, but outside the space bubble, many people see them both as greedy, tax-cheating, ego-driven boys building rockets in a dick-measuring contest. No doubt Elon's is "bigger" in the space community, and it is unlikely Jeff will catch up in the next decade, unless Elon trips.

SpaceX has a huge lead and is running faster than any of the competition, including all the big aerospace companies. To me, that is both fantastic and scary at the same time. Escaping gravity is not a simple maneuver and in the coming years it will be impossible to beat it safely every time. The private sector will have to answer to its customers for missteps that lead to bad outcomes. Only time will tell if they will be given the opportunity to correct their errors and continue as NASA has been allowed to do in the past.

Elon often stays in modest local housing when at his Boca Chica rocket ranch, but buying up much of the nearby land, including the town, has some locals in an uproar. Bezos is building a $500 million yacht but has pledged $10 billion to a new Earth fund aimed at funding efforts to fight inequities caused by climate change and decarbonize the economy. Neither have the best record with their former wives or employee practices, but with each in their fifties, there is plenty of time for their legacies and reputations to evolve if they set that intention.

Both men attribute their interest in space to reading science fiction as boys. The billionaires' mutual reverence for Heinlein and Asimov offers some insight into their apparently libertarian, male-centric worldviews. *The New York Times* ran a guest essay in late 2021 by Harvard history professor Jill Lepore, who highlights similarities in their beliefs to the technocracy movement of the early 1930s. Also inspired by science fiction, it espoused the conviction that technology and engineering could solve all political, social, and economic problems. Lepore points out that technocrats didn't trust democracy or politicians, capitalism or currency, and even objected to personal names.

As Elon Musk said on *Saturday Night Live* in 2021, "To anyone I've offended. I just want to say: I reinvented electric cars and I'm sending people to Mars in a rocket ship. Did you think I was going to be a chill and normal dude?" Well, he hasn't sent anyone to Mars yet, but he makes a fair point. Amassing that level of power and wealth is so unfathomable to most of us, it is easy to judge it as unseemly—we are wired differently. I find it interesting that there aren't yet examples of similarly wired women. Perhaps this can be attributed to a more collaborative, less competitive nature, but I can't help thinking the phenomena is at least partially environmental. The lyrics of Taylor Swift's song, "The Man" ring true to me: "I'd be a fearless leader, I'd be an alpha type, when everyone believes ya, what's that like?" I'd like to hear that song on the playlist for an astronaut's ride to the launchpad in her Tesla—it would certainly be on mine.

12.

THE VALUE PROPOSITION

POKING FUN AT WHAT IS BEING CALLED THE BILLIONAIRE SPACE RACE HAS become an art form, and I'll admit to laughing pretty hard at some of the parodies of my billionaire friends. Comparisons between them have attracted attention, but that is not the most consequential comparison. The true value proposition of our nation's human spaceflight activities requires an analysis between the projects managed internally and externally to NASA. Politicians on both sides of the aisle are missing the bigger picture.

It has become popular to rail against ruinous billionaire space ventures, while saying nothing about the nation's spending billions of tax dollars on lagging and inefficient NASA programs that undermine our international competitiveness. Such views would be analogous to complaints that the Wright Brothers flights at Kitty Hawk were frivolous, preferring to instead spend the public's money on the government's aviation exploits that at the time ended up literally being catapulted into the Potomac River.

Raising concerns over national policies that allow individuals to obtain massive wealth, and the negative environmental impact (in Earth and in space) of increased space activities is entirely reasonable and required. As elected leaders responsible for protecting the public welfare, setting more effective environmental and tax policies, and assuring that the government's space programs serve more universal purposes is directly in our national leaders' purview. I personally share these values and concerns, and I believe we must do a better job of establishing policies and regulations that incentivize behaviors that protect all our futures. It is us—our government—that is culpable. Targeting blame at the private space ventures instead of our lack of political fortitude is specious and misdirected.

NASA has had the mandate to reduce the cost of space transportation since the Nixon administration and that mandate has been consistently reinforced in space policy ever since. The Clinton administration's 1994 National Space Transportation Policy reiterated the goal clearly, saying, "NASA will assume the lead responsibility for research and development of next generation reusable systems." The government made little progress on its own because implementing the policy required letting go of existing expensive operational systems and dated self-serving goals. We still have not learned our lesson.

The human spaceflight programs that NASA is developing through traditional contracting methods, SLS and Orion, are tens of billions of dollars over budget and five years delayed. As many of us feared would happen, thousands of people have spent over a decade working on the system that through no fault of their own, was structured without any reusability or intention toward sustainability. The NASA Inspector General reported in November 2021 that its first four planned launches will each cost the government $4 billion. That's not even including the roughly $40 billion in sunk development costs. A recent Government Accountability Office report highlighted rampant, unwarranted award fees and uncovered that NASA has concealed billions of dollars in spending from Congress.

Meanwhile, both SpaceX and Blue Origin are developing privately funded heavy-lift reusable vehicles that will likely offer superior capability at a fraction of the cost. The Biden administration is now the third administration to ignore such realities, so the absurdity continues.

At least partially as an attempt to give SLS and Orion a mission, Vice President Pence announced in 2019 that NASA would land astronauts on the Moon by 2024. The administration named the program Artemis and marketed its intent to put the first woman on the Moon. Regarding the program's destination, President Trump contradicted himself and the Vice President several times, saying we should be going to Mars instead, since we've already been to the Moon. The space-industrial complex managed to ignore the diversion, wanting the contracts either destination would bring.

NASA's current plan is to develop a base on the Moon's south pole to explore lunar resources and prepare for a crewed mission to Mars. The rhetoric is not yet matched by funding, but Artemis's messaging was designed to appeal to all and has done so brilliantly within the space club. Estimated to cost an extra $30 billion over NASA's existing budget, the Trump administration's requests for an extra few billion dollars a year didn't appear to be sufficient, and what they did request was further cut over 50 percent by Congress. Unsurprisingly, rerunning a race to put the thirteenth person on the Moon, even if it finally includes a woman, has been a tougher sell to the public and some in Congress.

The rationale for returning to the Moon has been a mix of recreating a Cold-War competitive atmosphere, mostly against China, and hoping to inspire a new Artemis Generation. NASA is touting a set of unbinding principles it calls the Artemis Accords, that as of this writing have been signed by twelve other national space agencies. Although international collaboration has been a motivating justification for NASA's human spaceflight efforts over the last thirty years, Artemis is viewed as a US dominated activity. A handful of countries have agreed to provide hardware and a Canadian astronaut will join the first orbital flight. Most disquieting, Russia has been signaling that its post-ISS space efforts may align more with China than the US.

The Artemis program and schedule was hastily announced before plans were developed, but NASA is hard at work trying to make it a reality. The SLS rocket and Orion capsule, now scheduled for a first test flight this year, are only two of the elements needed to carry out the mission. The Agency is repurposing a spacecraft we initiated for the asteroid mission into a part of a human-tended space hub now called "Lunar Gateway." The NASA Inspector General reported its costs had increased significantly and that it won't be ready in time. NASA is configuring a work around for the first few missions. Another IG report released in the summer of 2021 said the space suits required for the mission would be delayed until at least 2025. They estimated that two suits would cost the taxpayers over a billion dollars, and although parts are being supplied by twenty-seven different contractors, NASA recently decided to bring the program "in house" to try to get it back on track.

In addition to the Gateway station and space suits; a lunar lander, rover, ground services equipment and experiments are in various stages of development. NASA's earthly infrastructure to support Artemis is nearly complete, but the costs are also out of this world. While SpaceX and Blue Origin have built or refurbished launch complexes for their similarly sized rockets at no cost to the public, NASA has spent a billion dollars to ready one for SLS.

An illustration of the two cultures can even be found in their different approaches to astronaut transportation on terra firma. NASA recently released a Request for Information (RFI) for an electric Crew Transportation Vehicle. The Agency is seeking to replace its former "Astrovan" in order to drive Artemis's four astronauts from the suit-up room to the launchpad, about four miles, scheduled to be done about once every two years (plus practice runs). The RFI calls for a vehicle with the capacity for a driver, four suited-up flight crew, three additional staff, room for six equipment bags, cooling units, and two additional cubic feet per passenger for miscellany. It also requires at least two large doors for entry/egress and an emergency exit. SpaceX manages to transport its four astronauts in their space suits, albeit less bulky than those used for Orion flights, on the exact same route with a couple of Tesla Model Xs.

• • •

When former Vice President Biden became the 2020 Democratic presidential nominee, I supported his candidacy and again found myself drafting space policy papers as a volunteer advisor. I helped write a congratulatory statement that the candidate released after the successful SpaceX Commercial Crew mission, drafted talking points, and agreed to participate in media availability in the lead up to Dragon's May launch. I was to join former Senator Bill Nelson, but a few days before the event an aide in the campaign's press office called—somewhat embarrassed—to tell me that Charlie Bolden was joining Bill Nelson for the media availability, so my assistance was no longer needed. Another campaign staffer later confirmed the change was made at Bill's request.

I remained committed to helping candidate Biden get elected and continued to support his space and climate policy efforts throughout the campaign. I received a call over the summer from one of Biden's early transition team organizers asking for my recommendations for the NASA review team. I was thrilled when Joe Biden won and I learned that some of my suggested candidates were named to the transition team. I enthusiastically provided them assistance when it was requested. As a consistent supporter of a White House National Space Council, I was especially pleased when the administration confirmed it would be retained and Vice President Harris would serve as its chair.

As usual, inauguration day came without a NASA Administrator being nominated. The transition team named a seasoned and well respected career employee as acting Administrator.

A cosmic series of events conspired to bait the new President into establishing his human spaceflight plans sooner than usual. Two weeks into the administration, Fox News reporter Kristin Fisher asked a question during the White House daily briefings about whether or not the Biden administration supported the Space Force that had been recently established by President Trump. When Press Secretary Jen Psaki reacted with a chuckle, making light of it before saying she'd have to look into it, the military space-industrial complex exploited the opportunity. Chastising the President and the press secretary for not taking such an important matter more seriously, they pulled out all the stops in an attempt to shame the White House into announcing its support for the new military service branch. It worked brilliantly.

When the press secretary confirmed the Space Force would be retained during the next day's press conference, the same reporter asked a follow-up question about the President's support for the Artemis program. Again, Jen Psaki didn't know what the reporter was referencing, but she knew not to make light and committed to getting her an answer. At the conclusion of the next day's press conference, Jen looked up from the podium for the reporter who'd asked the question as she pulled out her notes, reading what she had learned: that Artemis was a NASA program to take us back to the Moon and would even take a woman this time. She added that it sounded exciting, and she looked forward to sharing the information with her daughter. After one more round of questioning about whether or not the National

Space Council would be continued, the space community had what they wanted—they'd scored a hat trick.

It is a time-honored tradition for the press corps to ask questions in order to drive policy decisions, and the timing of the Fox News reporter's questions could not have been more well-played. To some, it appeared space had earned its place center stage as a priority policy issue being decided in the early days of an administration. As it became public that the reporter's parents were both astronauts, the insular nature of the community was criticized, but the check was already in the "win" column and Kristin Fisher had earned her allowance.

Having a personal interest in supporting space activities as a journalist or reporter on the beat is a pretty standard practice these days. Tweets and promotional videos disguised as "documentaries" about exciting and awe-inspiring space missions and launches are much more common than investigative pieces about how the government is spending the public's money. As in other niche issue areas, those drawn to careers reporting on specific topics are often also fans. Reporters who attempt more rigorous analysis do so at the risk of retaliation. The phenomenon is less improper in fully private space efforts and fields like entertainment and sports, since taxpayer dollars aren't involved, but it has the same impact. Journalists who write less than flattering stories about sports stars, celebrities or NASA Administrators risk losing their access to those they are paid to cover. The internet and social media have created a paradox for an independent fourth estate. The platforms facilitate greater availability of information, but the incentives are to boost eyeballs instead of knowledge—an issue related to the dearth of less polarized news sources in the mainstream media.

President Biden selected his long-time friend and former Senate colleague Bill Nelson to serve as his NASA Administrator in March. Nelson's nomination was universally acclaimed, partially fueled by the phenomenon outlined above, my own view being an exception. At the request of *Scientific American,* I penned an op-ed where I acknowledged the Senator's qualifications, but expressed concern that his nomination sent an unfortunate signal that the administration's goals might be outdated. My disappointment that Nelson would become the fourteenth consecutive man to lead the Agency was shared by many.

A woman who served on the Biden administration's NASA transition team, Pam Melroy, had been rumored to be in the running for the Administrator position. Pam is an astronaut, Air Force pilot, and one of only two women to command the Space Shuttle. She has degrees in physics, astronomy, Earth, and planetary sciences from Wellesley and MIT, and has held senior positions at Lockheed Martin, the FAA, and DARPA. She is well regarded in the community, and many of us hoped she would be President Biden's selection for Administrator.

A month after Administrator Senator Nelson's nomination, Pam was nominated for the deputy position. People familiar with their appointments have intimated that Bill asked for her nominating package to be held for release after his, in order to avoid any confusion over who was really in charge. When asked during his confirmation hearing what he would do to assure diversity was prioritized at NASA, Nelson responded that his deputy and CFO would be women, as if appointing a fourth female deputy and third female CFO was sufficient progress.

I expressed concern in my op-ed about Senator Nelson's having led the opposition to the President's, and Vice President's, budget proposal in 2010, which might indicate he was out of step with the Agency's most innovative and successful programs. Not surprisingly, the new NASA Administrator recalls his record differently. The seventy-nine-year-old is doing his best to wrap himself in the Commercial Crew flag. His former colleague Senator Kay Bailey Hutchison, no longer in the Senate, appeared as a special guest at his confirmation hearing, touting both of their long histories of championing the Commercial Crew concept and early program. Late night comics could have had their way with the "before" and "after" reels, but as usual, only those with a special interest took notice and most journalists stayed silently amused, not wanting to risk losing access to NASA's new leader.

Success has a thousand fathers, and failure is an orphan. I am elated that so many people are now supporting the transformational initiatives that they previously barely failed to sabotage. It was fully within Congress's right to question the new concept, but their unwillingness to get onboard with President Obama's proposal in favor of extending massive aerospace contracts, is a matter of public record. It has been humorous and gratifying to watch the program's former adversaries run to get on the fast-moving train headed in a different direction than the one for which they'd bought a ticket. Although he was not deeply involved with NASA at the time, Vice President Biden was always aboard the train and supported the policies the Obama administration prioritized.

President Biden's first NASA budget proposal included continued funding for Commercial Crew, SLS, Orion, and Artemis, with a 6 percent top-line increase over the previous year's budget. With trillions of dollars being spent on domestic stimulus, Administrator Nelson lobbied Congress to add $11 billion more to his budget, but as of this writing NASA has received about $1 billion, mostly directed to cover infrastructure improvements.

Administrator Nelson now says Artemis will land the first woman and first person of color on the Moon in 2025, and that "the Trump administration's target of a 2024 human landing was not grounded in technical feasibility." That ran counter to his previous messages, but as usual the media ignored the incongruity, understanding there will never be a shortage of excuses for delays.

A 2021 IG report estimates that the Artemis program will have cost the US taxpayer $96 billion through 2025, even though a landing on the Moon by then is not possible. NASA spent just twice that in comparative-year dollars for the entire Apollo program. NASA's Saturn V lunar rocket launched twelve missions, ten with crew, over five years. At best, the SLS will launch two to three times in five years. The Agency is ordering them through the decade and spending billions more on upgrades to take astronauts to Mars in the 2030s. Thankfully, while the dinosaurs devour the last of the leaves on the high treetops, the furry mammals have continued to evolve.

• • •

A few weeks before Senator Nelson was confirmed as Administrator, NASA selected SpaceX to build its lunar lander for the Artemis program.

SpaceX is leveraging the $2.9 billion fixed-price contract to accelerate the development of the Starship vehicle that it has been building for many years at their own cost. SpaceX's selection opens up the opportunity to eventually transition away from the expensive government-owned systems that are still being developed for Artemis. If successful, Starship alone could perform the entire Artemis mission without SLS, Orion, or the Lunar Gateway, at significantly reduced cost and increased capability. The shift to a more sustainable architecture for human space exploration again feels in reach.

On the day of the announcement, one of my former NASA colleagues who still works at the Agency sent a private message reminding me that of all the major human spaceflight contract awards in the last decade, only one was a cost plus FAR contract won by a traditional defense contractor. He wrote that they "used to have to whisper the phrase 'funded Space Act Agreement' around NASA. Now, everyone whispers 'cost plus' like it's some form of cancer. What a transformation!"

NASA had hoped their budget would be enough to select two lunar lander winners, but the appropriation from Congress would barely cover one. Without funding a competitor, the Agency has put a lot on SpaceX's shoulders. Although the losing teams' proposals were double the cost, they filed formal protests, and after their initial protest failed, Blue Origin took NASA to court and lost again. NASA plans to offer an on-ramp for additional services, so there will likely be more opportunities for Blue Origin and others to compete on the lunar initiative.

Many of the same senators and congressmen who balked at NASA's selecting two competitors for the Commercial Crew program now claim it to be a requirement for the lunar lander. The Senate passed legislation demanding NASA award a second contract, without giving them the money to pay for it. The unfunded mandate is, as usual, more parochial than partisan.

The Human Landing System (HLS) contract doesn't fit the usual standards for fixed-price contracts, since their use is best suited to mature technical programs where requirements are well-known. Ironically, one of the reasons a fixed-price award is even an option for the HLS, is because of SpaceX and Blue Origin's willingness to share the government's costs and risks. If the program had been procured through a more typical cost-plus contract, no one would have batted an eye over the selection of a single contractor. Bats are being swung not because two awards weren't given out, but because the winner is SpaceX.

Congress should be ecstatic with the company's demonstrated cost and performance, but the forces opposed to its now obvious ability to run the table still exercise power. The reality is, that without SpaceX contributing its own resources to develop Starship, the government wouldn't have the money for even a single lunar lander, and Artemis wouldn't be much more than a great name for a human spaceflight program.

The tendency to project one's views of the most visible individuals involved in advancing space development, onto the activity itself, is common. But whether we personally like the billionaire space titans as individuals is beside the point. By all accounts, they are following established laws, and instead of investing in space companies, they could be spending all of their money on creature comforts that do little for our national economy.

Now some of the space pirates' loyalties are even being divided. There is less focus on keeping the ball moving down the field to sustain progress for all. It is too early to claim victory just because we moved the ball into the red zone. Americans are known for loving a good competition and this one is attracting more attention than the Super Bowl, so sides are being taken. But the traditional players haven't retired; they are writing new plays while enjoying *and fueling* the fratricide. In my view, we still need to keep our eye on the ball in order to assure sustainable progress.

When the Obama administration gave in to the overwhelming assault from the self-interested parties to build its own heavy-lift rocket, I was vilified for opposing the plan. It felt like my character and patriotism were called into question because I pointed out that major cost increases and delays were inevitable and that the government shouldn't be building its own rocket because it would compete with the private sector. The space pirates agreed, and many spoke out, but more importantly—they didn't give up.

I figured SpaceX could get astronauts to the ISS—and maybe even develop the Falcon Heavy—before the SLS flew, for about 10 percent of the money. Those were not the odds the established bookies would have given, but lots of people would have joined me (mostly privately) in placing those bets and long since banked their winnings. This is only the beginning of the story.

Both Blue Origin and SpaceX have been building heavy lift reusable rockets at their own expense and are already nipping at the heels of the government's throwaway rocket that receives billions of our tax dollars. I would not have allowed myself to hope for, much less predict, such progress. Blue Origin is a few years later than its estimated timing in launching its New Glenn rocket, but since it is internally funded, missing self-imposed deadlines is its own concern. SpaceX is hard at work at Starbase, now utilizing NASA's lunar-lander funding to accelerate development of its Starship. Progress appears to be swift, but looks can be deceiving.

If SpaceX is able to bring the program online in the next few years, the game will change again. Starship is radically different than everything that has come before. Starship is even larger than the SLS, but its size isn't its most profound differentiator. The rocket is designed to launch a hundred people and land them anywhere back on Earth—or on the Moon or on Mars—ready to be fueled and launched again.

Each element of the rocket and spacecraft are already being flight tested in the skies of Texas. Sometimes the hardware and software perform as expected and sometimes they don't, but either way the lessons are woven into the next version of the vehicle. It is hard to grasp what the impact would be of an operational Starship system, but it would be revolutionary, and therefore blood is already being spilled.

The stakeholders who brought us SLS and Orion are heavily invested in protecting them. Their standard critique of Starship is that most of its flight tests have ended without an intact rocket. Such criticism ignore the fact that no orbital rockets have ever ended with an intact rocket, since none have even attempted powered vertical landings. The concept of reusability is still lost on some in the space community.

Charlie Bolden recently said that "if we lost rockets at the rate Elon Musk loses his big Starship, NASA would have been out of business. Congress would have shut us down." SpaceX's rockets aren't being "lost." They are test articles. This is the kind of thinking that has trapped NASA and limited it to traditional, expensive programs at the expense of a more measured, iterative response to risk. Now calling himself a "huge fan of SpaceX, but a huge skeptic of Starship," Charlie said in an interview in late 2021 that the difficulty for him is the fact that it is so big, so massive. He added: "If Neil Armstrong was alive today to talk to him, he would probably say, 'That is the dumbest thing I've ever heard.'" Perhaps the time has passed for us to design policies and programs based on what retired astronauts may or may not think.

Imagine if ships, trains, cars, or airplanes had remained in the government's control and the vehicles were initially designed for a single use? No mode of transportation would have led to much progress until someone came along and figured out a way to make them reusable. The value proposition for space activities, like it has for all other means of transiting new environments, is finally tipping the scales in the right direction. Financial investors are rushing to fuel potential shooting stars.

It took non-vested interests with the resources to take on the space-industrial complex to jump-start the transition to a new space age. Thanks to a handful of space pirates, billionaires, and bureaucrats willing to stand up to the system of patronage, progress is now being realized. A program that was scorned by the establishment when it was introduced is now using innovative, reusable, private-sector-driven technologies to provide space transportation at a fraction of the cost of past government owned and operated programs—just imagine what else is possible.

• • •

Humanity's first leaps off the planet were begun as a competition with the Soviet Union that incentivized speed over endurance. While this early motivation may have impeded sustained progress, its intent was never to say "this far and no further." Ships sank, airplanes crashed, and the countries that gave up turned inward, but civilization has always continued to evolve. Exploration is, after all, driven by our quest to survive. Nations who learned to utilize the oceans and the atmosphere for public benefit—whether out of fear, greed, or glory—thrived.

Exploring space has allowed us to begin to understand the mysteries of the universe, including how life began and whether it exists elsewhere in any form. NASA missions have opened our eyes to what lies beyond our atmosphere, as well as on Earth below. A reformed NASA can be the rising tide that lifts all boats.

In an evolutionary sense, our tide is rising, but at rates faster than much of nature is able to adapt. Evolution teaches that animal life crawled out of the ocean nearly 400 million years ago, and before that, it may have come to planet Earth from space. Gene Roddenberry once told me that even though our emergence from the sea was more recent, he believed our pull to space was stronger, since we must eventually return to sustain humanity. Gene's remark has resonated with me ever since.

NASA is a national asset that if properly reformed can continue to make meaningful contributions to sustaining humanity on Earth and eventually beyond. Recent human activities are changing our home planet in ways that aren't easy to see from our own backyards. To understand what is happening here, we need to see ourselves from a new perspective, one that shows how we are connected—7.7 billion people and 8.7 million species—to the only living planet in our own galaxy or in the known universe. It is only from this perspective that we can fully understand what can be done to allow planet Earth to remain a vital home to future generations.

The Industrial Age fueled population expansion across the globe and our first steps beyond. In the Digital Age, we now gather and instantaneously access massive amounts of data from space that inform Earth systems models, revealing how unprecedented amounts of greenhouse gasses being released into our atmosphere are causing a climate crisis that threatens our existence.

Temperatures in the atmosphere, land, and oceans are increasing; glaciers are melting, and seas are rising. These changes are fueling extreme weather events, catastrophic storms, flooding, and droughts that affect every aspect of our environment—air quality, water availability, food supply, biodiversity, and disease. All of life as we know it is under stress. Data shows that over the next few decades—a blink of an eye in Earth's story—the damage we have caused will accelerate uncontrollably, making it even more difficult, if not impossible, for us to reverse it. We are facing the tipping point for human life on our home planet. What we do now will determine the rest of our story.

Thanks to advancements in space development, we live in a moment in history when scientific and technological progress allow us to understand the negative impacts of previous inventions, offering us a rare and fleeting opportunity to recover. Armed with the knowledge of what is happening and why, our new perspective offers solutions. Global, high-fidelity, verifiable satellite data can be utilized to validate and enforce policies and treaties that reduce greenhouse gas emissions. Improved sensor technologies, data accessibility, and distribution can provide critical, timely information to more precisely measure, model, predict, and adapt to the climate crisis, limiting human suffering. NASA has the experience, organizational credibility, and expertise to contribute more to this effort.

NASA can establish programs that address these challenges within its existing mandate. After all, the Agency was not created to do something again. It was created to push the limits of human understanding and help the nation solve big, impossible problems that can benefit from scientific and technological advancements located off Earth's surface.

Sixty-one years ago, NASA was challenged to reach farther and, in successfully achieving this goal, provided a new perspective of ourselves and our beautiful, fragile home planet. Apollo 8 astronaut Bill Anders, who took the *Earthrise* photo capturing Earth as it rose behind the Moon, said, "We came all this way to explore the Moon, and the most important thing is that we discovered the Earth."

President Kennedy's speech that set the stage for Apollo explained the challenge poetically: "We set sail on this new sea because there is new knowledge to be gained, and new rights to be won, and they must be won and used for the progress of all people."

Future space voyages are on a trajectory to again lift all people by providing new knowledge and resources, while more fully utilizing atmospheric and space-based science and technology to address society's immediate challenges.

Investments in space activities have led to advancements that now allow us to tap the collective genius of humanity to find solutions to the previously insurmountable problems we face today. Expanding our presence beyond our own planet isn't just about escaping gravity. It can be part of a larger strategy that addresses the gravity of our situation. The stakes have never been higher, and although failure is an option, it is not one we should risk.

We need leaders who will put forward effective policies and programs that tackle current societal threats and who are willing to stand up to rooted, powerful special interests. Dismantling existing policies, entrenched bureaucracies, and industries is never popular with the institutions and people most directly affected. Our future depends on recognizing that the proper role of government is to support the greater good.

My generation grew up seeing the United States as a leader at least partly because of our ability to explore space. It is fitting that the reforms at NASA that have incentivized innovative and effective programs can now light a path for other stalled government activities. Holding on to the past is robbing the next generations of their chance for a healthy and prosperous future.

The most important lesson we learned from our first forays beyond our thin atmosphere was that we are in this together. We overcame gravity by working collectively toward an aligned goal. That same force must now allow us to overcome political and policy differences that are driven by circumstances such as what we look like, where we live, and who we love.

Keeping our shared end state in mind can give us the ability to use our knowledge for the most meaningful of purposes. Our living, breathing home planet is our cradle, and although we need to eventually leave it to survive, sustaining ourselves on Earth is a giant leap that requires our united determination. As Carl Sagan observed nearly thirty years ago, "There is no hint that help will come from elsewhere to save us from ourselves."

EPILOGUE

STALLED SPACE DEVELOPMENT HAS NOT BEEN THE ONLY ADVERSE RESULT of federal policies that extend costly, unnecessary programs at the expense of progress. President Eisenhower's fears about the negative effects of the runaway power of the military-industrial complex have been realized. But they need not continue. NASA's small investments in private sector developments have shown how others can out-compete traditional defense companies in the most challenging of arenas: human spaceflight. The point is not to have existing companies fail, but to incentivize them to become better through real competition. IBM didn't cease to exist when Microsoft came on the scene, but they were forced to improve.

As with NASA, spending for large programs in congressional districts where federal and industry jobs already exist is often prioritized over effectiveness. Protecting current military spending and infrastructure perpetuates outdated programs focused on winning the last war instead of the next. Our nation's lack of preparedness for COVID should be a wake-up call to us all. The nonpartisan government watchdog Project on Government Oversight sums up what many of us are now thinking:

> It's certainly one of the stranger phenomena of our era: after twenty years of endless war in which trillions of dollars were spent and hundreds of thousands died on all sides without the US military achieving anything approaching victory, the Pentagon continues to be funded at staggering levels, while funding to deal with the greatest threats to our safety and national security—from the pandemic to climate change to white supremacy—proves woefully inadequate. In good times and

bad, the US military and the "industrial complex" that surrounds it, which President Dwight D. Eisenhower first warned us about in 1961, continue to maintain a central role in Washington, even though they're remarkably irrelevant to the biggest challenges facing our democracy.

Security risks to US citizens and to all of humanity have evolved more quickly than the systems we have in place to address them. As is the case with our civil space programs, military programs need to be reformed and retooled to address current threats to a peaceful and prosperous nation. President Eisenhower, a former five-star general, cut the defense budget by 27 percent during his time in office and argued that the "world in arms is not spending money alone. It is spending the sweat of its laborers, the genius of its scientists, the hopes of its children."

Over the last decade, nonpartisan and bipartisan reviews such as the Bowles-Simpson fiscal commission have recommended Department of Defense efficiencies that could be redirected to more beneficial and effective public programs. Applying right-to-left thinking would result in dramatic investment shifts and improvement in public health and safety as well as our global national security strategy. The government reform group Public Citizen tweeted in January 2021, "If you're spending $740,000,000,000 annually on 'defense' but fascists dressed for the renaissance fair can still storm the Capitol as they please, maybe it's time to rethink national security?" Unless this paradigm changes, the incentive to keep repeating the cycle will continue to support the giant self-licking ice cream cone, funding programs disconnected from modern society.

Realigning national policies and spending to develop vaccines have shown the latent power of the government to leverage private sector capabilities, fueling scientific and technological advances to unite to solve global problems. It has also exposed fault lines. Government policies should incentivize individuals, nonprofit organizations, and corporations of all sizes to drive innovations that will respond to today's challenges, instead of spending massive public resources to prop up outdated infrastructure and weapons systems aimed at meeting former threats and fighting past enemies.

Neither the incentives nor the consequences that exist in the government today are sufficient to fulfill the Founders' purposes of providing for the welfare and security of society. Citizens self-select the information and news we hear, which has reduced our ability to think for ourselves. This phenomena undermines a fundamental tenant of healthy democracies—an informed citizenry. I don't know if term limits, public campaign financing, better regulation of fake news, limiting wealth, monetizing carbon and other GHG emissions, or realigning the tax code are the answer, but all seem worthy of serious consideration. I'm not an expert, but from my perch, we suffer from a lack of attention to the bigger picture. We could use more people like my late mid-Michigan farmer grandfather and uncle, whose public service aspirations were to help their neighbors.

Agreeing on the basic purposes of government and creating policies, institutions, and federal budgets to address the nation's current reality is what the framers of our Constitution faced 250 years ago. Our job today is harder. Renovating a house requires different skills than building one anew. Hard choices must be made in order to replace our clogged pipes and rotted wood with contemporary tools and materials that will shore up our foundation to withstand another few centuries. We have the experience and knowledge to succeed if we can move beyond past divisions and embrace each other from a new perspective.

AUTHOR'S NOTE

NASA OCCUPIES A REVERED PLACE IN OUR NATION'S PSYCHE AND IS THUS frequently memorialized. The intersection of politics and spaceflight has been depicted by Hollywood and dissected in academic settings, but the true essence of the relationship has been shrouded in mystery. The importance of the transformation in human spaceflight that occurred over the last decade has led many new journalists and historians to opine on the subject.

Memoirs by former NASA political leaders are conspicuous in their absence, which has led to much speculation and secondhand interpretation. I wrote *Escaping Gravity* to offer my own perspective. My intention was to demystify and elevate the politics of US human spaceflight, and provide an example of how other government programs could be improved.

I began contemplating the idea of writing a book on this topic shortly after leaving NASA in 2013. I created numerous outlines and occasionally jotted down notes without making much progress. Collaboration has been a central tenet of my career, and I realized the project was suffering from that missing element. I searched for a journalist to fill gaps in my knowledge of current developments in the field, and in a 2019 version of a cold call, I DM'd @thesheetztweetz—CNBC reporter Michael Sheetz—that spring. Although we had never spoken before, Michael was quick to reply and expressed his interest in the pursuit.

We met occasionally over the next year in New York City and DC and began a collaboration. The SpaceX Demo-2 launch inspired us to announce our plans, and CNBC published a portion of an early chapter in May 2020. The pandemic made working in person challenging and as the narrative evolved, it became my story to tell. Even so, Michael's insights and early contributions were numerous. *Escaping Gravity* benefited greatly from his involvement and I am deeply thankful for Michael's guidance and friendship.

I am also grateful to the many past and current colleagues who generously offered their time for interviews, random inquiries and reviews of the draft manuscript, including: Royce Dalby, Rebecca Spyke Keiser, George Whitesides, Beth Robinson, Casey Handmer, Phil McAlister, Rich Leshner, Dan Hammer, Will Pomerantz, James Muncy, David Weaver, Laurie Leshin, Phil Larson, Jeff Manber, Mark Albrecht, Courtney Stadd, Elise Nelson, Dan Goldin, and Alan Ladwig. To my publisher, Scott Waxman of Diversion Books and editor Keith Wallman, thank you for your willingness to take a chance on me and my story. Thanks also to Diversion's Evan Phail, who held my hand and shepherded me through the mysterious world of publishing.

Thanking one's family is a time-honored tradition for authors, but I now understand how well it is deserved. Dave, Wes, and Mitch Brandt read and edited numerous versions of the book and provided valuable input. They somehow learned how to manage their lives without me, while putting up with my need to incessantly converse on the topic for two years. Dave did this while filling in for my absence from social activities, parenting responsibilities, and dog walking (although perhaps that is not a new phenomenon). My mother, sister, brother-in-law, and numerous friends read draft manuscripts, offering constructive critiques and encouragement. I am eternally grateful for my family and friend's love and support.

Writing a book is a huge undertaking for anyone, but it is a special challenge for those of us who are not professional writers. Balancing policy details with personal stories meant sometimes glossing over the finer points of issues and people near and dear to my heart. These compromises led to only a handful of the individuals who deserve credit being identified and fewer technical specifics, but hopefully contributed to a more accessible narrative.

People in power often prefer to operate under the veil of secrecy, but public servants' actions should withstand public scrutiny. Those I describe as working against what I thought were needed reforms are not bad people. In my view, they are products of a system where their professional status led them to assume privilege. The halls of power were filled with others who looked and acted like them, which reinforced their beliefs and behavior. To a person, they have made many positive contributions to the nation and space program throughout their careers. My retellings of our interactions are not meant to reflect negatively on their intentions or their other accomplishments. Marrying colleagues' positive reputations and numerous good-hearted deeds with what I came to experience was confounding. I air what I saw as misdeeds not out of spite but with a belief that sunlight can be a powerful disinfectant.

I have utilized source material throughout the book as much as possible, but where others' recollections or interpretations of the same events or conversations differ, I apologize for any lapses in my own account. I requested and received permission to write about my shared experiences with the people I named in the above review credits, as well as from Steve Isakowitz, Fisk Johnson, Peter Diamandis, and Mary Ellen Weber. There were others who provided information on background but have asked not to be named. I reached out to Lance Bass, but wasn't able to connect. If he reads the book, I hope the pages on Astromom and the Basstronaut give him the same smile they gave me to recall.

My recollections of conversations with the most famous people in my story—President Obama, Tom Hanks, Elon Musk, Jeff Bezos, Richard Branson, and others—are from my own memory. I haven't talked to them recently, but given their exalted status, our conversations were memorable. That is likely much less so for them, but I hope my characterizations ring true if they read the book.

NASA astronauts are all exceptionally brave, driven, smart, technically proficient, and physiologically high-caliber individuals. They are paid a decent, but not extravagant amount of the public's money to do their jobs, which only rarely includes going to space. They are not a monolithic group, but for simplicity are sometimes referred to as such in *Escaping Gravity*.

Having famous astronauts as close colleagues and friends hasn't made me immune to hero worship. Finding myself in the crosshairs of public icons was intimidating. The super powers society ascribe to astronauts made holding positions opposed to theirs harrowing. Evaluating how to explain my *less than positive* encounters with the hero astronauts depicted in the book was agonizing. If the meaning of my story could have been conveyed without sharing this context or our conflicts, I would gladly have done so. Just as not every professional baseball player would make a good general manager and doctors aren't all equipped to run hospitals, not all astronauts are suited for every profession. NASA astronauts are trained to do complex and precise physical and technical tasks, in confined environments, with few people, under extreme stress. These are their super powers. The nation has few remaining heroes and NASA astronauts are in a class of their own.

As more and more people travel to and through space, the mystique of the astronauts will eventually recede. Ship captains and commercial airline pilots no longer enjoy the public recognition they received in the early years of seafaring or the Jet Age. Similarly, history will capture the names of those who opened the new frontier and the heroism they displayed. It is understandable that some astronauts prefer to keep the club small and feel protective of their title. I decided to refer to anyone who has chosen to risk their life by allowing themselves to be hurled above fifty miles, the FAA's definition of "space," as astronauts. The word was created by science fiction writers long before NASA was conceived. First appearing in written form in the 1920s, a combination of the Greek words "astron" (star) and "nautes" (sailor), it conjures an astronaut as a sailor among stars. Titles don't typically signify sameness. Sailors can traverse vast oceans or small lakes, and doctors don't all operate on people. Even within the NASA astronaut corps, the range of training and experience vary greatly. Those who walked on the Moon, have an elevated status. Pilot astronauts sometimes dismissed mission and payload specialists as less than full astronauts in the early days of the Shuttle.

Buckminster Fuller said, "We are all astronauts on a little spaceship we call Earth." This is the message of *Escaping Gravity*. Instead of focusing on our differences, we would do well to focus on our similarities. We are all like the crew of a large ship, and people have to work together in order to keep the planet functioning.

I've taken liberties with the "space pirate" moniker and apologize to anyone who would have preferred not to be associated with this label or who feel I've mischaracterized their role. An earlier book title the publisher considered was *Space Pirates*, and my intention in telling this story is to honor their quest. Like all worthwhile endeavors, the transformative progress I describe in *Escaping Gravity* was only possible because of a multitude of individuals and organizations. A list of those deserving credit would fill many pages, but they are truly the people responsible for launching a new space age.

This is not meant to be an academic dissertation or historical reporting of the era. It is a memoir. I've undoubtedly fallen short of capturing all the meaningful elements and events that contributed to the progress we achieved. Any omissions are my own and are not intended to be dismissive of their importance. NASA is a crown jewel of our nation, and it is my deepest hope that the perspective offered here will allow it to continue to advance and benefit humanity in the future.

ACRONYMS

ADA – Antideficiency Act
AMROC – American Rocket Company
AMS – Alpha Magnetic Spectrometer
ARM – Asteroid Redirect Mission
ASAP – Aerospace Safety Advisory Panel
ASCANS – Astronaut Candidates
ATK – Alliant Techsystems, later merged with Orbital Sciences Corp
BEAM – Bigelow Expanding Activity Module
CAIB – Columbia Accident Investigation Board
CCiCap – Commercial Crew Integrated Capability
CJ – Congressional Justification
COTS – Commercial Orbital Transportation Services
COTS-D – Designation for crew services in addition to cargo
COMSAT – Communications Satellite Corporation
DARPA – Defense Advanced Research Projects Agency
ENIAC – Electric, Numerical, Integrator and Computer
ESA – European Space Agency
EELV – Evolved, Expendable Launch Vehicle
EOP – Executive Office of the President
FAR – Federal Acquisition Regulation
FAA – Federal Aviation Administration
GAO – Government Accountability Office
HLS – Human Landing System
IBMP – Institute of Physical and Biological Problems (Russian)
ICBM – Inter-Continental Ballistic Missile
IG – Inspector General
INTELSAT – International Telecommunications Satellite Organization

ISS – International Space Station
JPL – Jet Propulsion Laboratory
JSC – Johnson Space Center
KSC – Kennedy Space Center
L5 – Society founded by Dr. Gerard O'Neill
LOCV – Loss of Crew and Vehicle
LEO – Low Earth Orbit
MSFC – Marshall Space Flight Center
MSL – Mars Science Laboratory
NIH – National Institutes of Health
NASA – National Aeronautics and Space Administration
NRO – National Reconnaissance Office
NSI – National Space Institute
NSS – National Space Society
OMB – Office of Management and Budget
OMEGA – Offshore Membrane Enclosures for Growing Algae
OSTP – Office of Science and Technology Policy
POTUS – President of the United States
RFI – Request for Information
RFP – Request for Proposals
RLV – Reusable Launch Vehicle
RSA – Russian Space Agency, later known as Roscosmos
SCIF – Secret Compartmentalized Information Facility
SEC – Securities and Exchange Commission
SEI – Space Exploration Initiative
SFOC – Space Flight Operations Facility
SLS – Space Launch System
SpaceX – Space Exploration Technologies Corporation
STS-107 – Designation of ill-fated Space Shuttle Columbia mission
STS-135 – Designation of the final Space Shuttle mission
TRW – Aerospace company acquired by Northrop Grumman in 2002
ULA – United Launch Alliance
UNIVAC – Universal Automatic Computer
USA – United Space Alliance
VASMIR – Variable Specific Impulse Magnetoplasma Rocket
X-33 – NASA flight demonstration program awarded to Lockheed Martin

SOURCES

The sources are listed in the order in which they were used in support of the text within each chapter.

1. GAME CHANGER

"President Nixon's 1972 Announcement on the Space Shuttle." The Statement by President Nixon, 5 January 1972. history.nasa.gov (https://history.nasa.gov/stsnixon.htm)

Space News Staff. "Obama Adds Three More to NASA Transition Team." Space.com. November 25, 2008. (https://www.space.com/6161-obama-adds-nasa-transition-team.html)

Columbia Accident Investigation Board Final Report. "On Feb. 1, 2003, Shuttle Columbia Was Lost During Its Return to Earth. Investigators Have Found the Cause." (https://govinfo.library.unt.edu/caib/default.html)

"Overview: Ares I Crew Launch Vehicle." NASA. (https://www.nasa.gov/mission_pages/constellation/ares/aresI_old.html)

Atkinson, Nancy. "Obama to Re-examine Constellation Program." Universe Today. May 5, 2009. (https://www.universetoday.com/30384/obama-to-re-examine-constellation-program/)

"U.S. Announces Review of Human Space Flight Plans." The White House Archives. May 7, 2009. Obamawhitehouse.archives.gov (https://obamawhitehouse.archives.gov/the-press-office/2015/11/16/us-announces-review-human-space-flight-plans)

Review of U.S. Human Spaceflight Plans Committee. "Seeking a Human Spaceflight Program Worth of a Great Nation." October 2009. NASA. (https://www.nasa.gov/pdf/396093main_HSF_Cmte_FinalReport.pdf)

NASA Fiscal Year 2011 Budget Estimate. NASA. (https://www.nasa.gov/news/budget/2011.html)

Messier, Doug. "Attacks on Lori Garver Backfiring." Parabolic Arc. February 25, 2010. (http://www.parabolicarc.com/2010/02/25/attacks-lori-garver-backfiring/comment-page-1/)

News Release from US Senator Richard Shelby (R-Ala.). "Shelby: NASA Budget Begins Death March for U.S. Human Space Flight." February 1, 2010.

(https://www.shelby.senate.gov/public/index.cfm/newsreleases?ID=8A4B0876-802A-23AD-43F9-B1A7757AD978)

Jones, Richard M. "Senator Nelson on NASA's FY 2011 Budget Request." FYI: Science Policy News from AIP. American Institute of Physics. February 18, 2010. (https://www.aip.org/fyi/2010/senator-nelson-nasa's-fy-2011-budget-request)

Pasztor, Andy. "Senators Vow to Fight NASA Outsource Plan." *The Wall Street Journal.* February 24, 2010. (https://www.wsj.com/articles/SB10001424052748704240004575085900217022956)

U.S. Senate Committee on Commerce, Science, & Transportation Hearings. "Challenges and Opportunities in the NASA FY 2011 Budget Proposal." Webcast. February 24, 2010. (https://www.commerce.senate.gov/2010/2/challenges-and-opportunities-in-the-nasa-fy-2011-budget-proposal)

Klamper, Amy. "Sen. Nelson Floats Alternate Use for NASA Commercial Crew Money." *SpaceNews.* March 19, 2010. (https://spacenews.com/sen-nelson-floats-alternate-use-nasa-commercial-crew-money/)

Klamper, Amy. "Obama's NASA Overhaul Encounters Continued Congressional Resistance." *SpaceNews.* April 23, 2010. (https://spacenews.com/obamas-nasa%E2%80%82overhaul-encounters-continued-congressional-resistance/)

Sutter, John D. "Obama Budget Would Cut Moon Exploration Program." CNN. March 15, 2010. (http://www.cnn.com/2010/TECH/space/02/01/nasa.budget.moon/index.html)

Maliq, Tarik. "Neil Armstrong Blasts Obama's Plan for NASA." *The Christian Science Monitor.* May 14, 2010. (https://www.csmonitor.com/Science/2010/0514/Neil-Armstrong-blasts-Obama-s-plan-for-NASA)

O'Keefe, Ed and Marc Kaufman. "Astronauts Neil Armstrong, Eugene Cernan Oppose Obama's Spaceflight Plans." *The Washington Post.* May 12, 2010. (https://www.washingtonpost.com/wp-dyn/content/article/2010/05/12/AR2010051204404.html)

Armstrong, Neil. "Future Space Opportunities Are the President's Call." *Wall Street Journal.* December 27, 2008. (https://www.wsj.com/articles/SB123033959209636593)

NASA. "Report of the Space Task Group, 1969." https://history.nasa.gov/taskgrp.html (https://history.nasa.gov/taskgrp.html)

Space Foundation Research & Analysis. "Space Data Insights: NASA Budget, 1959-2020." The Space Report Online. (https://www.thespacereport.org/uncategorized/space-data-insights-nasa-budget-1959-2020/)

Editors. "President Nixon Launches Space Shuttle Program." HISTORY. November 16, 2009. (https://www.history.com/this-day-in-history/nixon-launches-the-space-shuttle-program History.com)

H.R. 3942 (98th): Commercial Space Launch Act. "Commercial Space Launch Act of 1984." October 30, 1984. (https://www.govtrack.us/congress/bills/98/hr3942/text)

President Reagan's Statement on the International Space Station. "Excerpts of President Reagan's State of the Union Address, 25 January 1984." NASA. (https://history.nasa.gov/reagan84.htm)

White, Frank. *The Overview Effect: Space Exploration and Human Evolution*. Multiverse Publishing LLC. 1987.

2. STAR STRUCK

Krauss, Clifford. "House Retains Space Station in a Close Vote." *The New York Times*. June 24, 1993. (https://www.nytimes.com/1993/06/24/us/house-retains-space-station-in-a-close-vote.html)

Myers, Laura. "Lovell Gets Medal of Honor, Confesses Costner His First Pick to Play Him." Associated Press. July 26, 1995. (https://apnews.com/article/ea59ee78e4591ef1917ea1afa780dba6)

Daalder, Ivo H. "Decision to Intervene: How the War in Bosnia Ended." Brookings. December 1, 1998. (https://www.brookings.edu/articles/decision-to-intervene-how-the-war-in-bosnia-ended/)

Dick, Steven (editor). *NASA 50th Anniversary Proceedings: NASA's First 50 Years: Historical Perspectives*. NASA. p. 166.

Cowen, Robert C. "Bush, Dukakis on Space." *The Christian Science Monitor*. September 23, 1998. (https://www.csmonitor.com/1988/0923/a1spac5.html)

Anderson, Gregory. "A Few Words with Newt Gingrich." The Space Review. May 15, 2006. (https://www.thespacereview.com/article/623/1)

Foust, Jeff. "Gingrich Ends His Campaign, But Not His Interest in Space." Space Politics. May 3, 2012. (http://www.spacepolitics.com/2012/05/03/gingrich-ends-his-campaign-but-not-his-interest-in-space/)

Quayle, Dan. *Standing Firm*. Harper Collins Zondervan. 1994. pp. 179–181.

Broad, William J. "Lab Offers to Develop an Inflatable Space Base." *The New York Times*. November 14, 1989. (https://www.nytimes.com/1989/11/14/science/lab-offers-to-develop-an-inflatable-space-base.html)

NASA Internal Report. "Report of the 90-Day Study of Human Exploration of the Moon and Mars." NASA. November 1989. (https://history.nasa.gov/90_day_study.pdf)

Lori Garver, Executive Director, National Space Society, October 26, 1990 testimony to the Advisory Committee on the Future of the U.S. Space Program. NASA History Division. December 1990. (https://space.nss.org/wp-content/uploads/Advisory-Committee-On-the-Future-of-the-US-Space-Program-Augustine-Report-1990.pdf)

Albrecht, Mark J. *Falling Back to Earth: A First Hand Account of the Great Space Race and The End of the Cold War*. New Media Books. 2011. p. xv.

Gerstenzang, James. "Bush Denounces NASA Fund Cuts : Space: The President Says Exploration Programs Cannot Wait Until All of the Nation's Social Ills Are Solved. He Also Stumps for Helms in North Carolina." *Los Angeles Times*. June 21, 1990. (https://www.latimes.com/archives/la-xpm-1990-06-21-mn-156-story.html)

Sawyer, Kathy. "Truly Fired as NASA Chief, Apparently at Quayle Behest." *The Washington Post*. February 13, 1992. (https://www.washingtonpost.com/archive/politics/1992/02/13/truly-fired-as-nasa-chief-apparently-at-quayle-behest/bc7cc6cc-1799-4435-8550-e879d81dcff1/)

Pasternak, Judy. "Bush Nominates TRW Executive to Head NASA." *Los Angeles Times.* March 12, 1992. (https://www.latimes.com/archives/la-xpm-1992-03-12-mn-5289-story.html)

Telephone interview with Dan Goldin, June 20, 2020.

Dunn, Sarah (editor). "U.S.-Soviet Cooperation in Outer Space, Part 2: From Shuttle-Mir to the International Space Station." National Security Archive. The George Washington University. May 7, 2021. (https://nsarchive.gwu.edu/briefing-book/russia-programs/2021-05-07/us-soviet-cooperation-outer-space-part-2)

Oberg, James. *Star-Crossed Orbits: Inside the U.S.-Russian Space Alliance.* McGraw-Hill. 2001.

Wilford, John Noble. "NASA Loses Communication With Mars Observer." *The New York Times.* August 23, 1993. (https://www.nytimes.com/1993/08/23/us/nasa-loses-communication-with-mars-observer.html)

Cappiello, Janet L. "Hubble Error Due to Upside-down Measuring Rod." Associated Press. September 14, 1990. (https://apnews.com/article/a080cf57761942b3a6837eb87b088bc5)

Leary, Warren E. "NASA Is Urged to Push Space Commercialization." *The New York Times.* February 8, 1997. (https://www.nytimes.com/1997/02/08/us/nasa-is-urged-to-push-space-commercialization.html)

Bell, Julie. "NASA to License Its Space Simulator Today." *The Baltimore Sun.* September 14, 2000. (https://www.baltimoresun.com/news/bs-xpm-2000-09-14-0009140148-story.html)

Burke, Michael. "Medical Research Investment Takes Off for Fisk Johnson." *Journal Times.* January 23, 2002. (https://journaltimes.com/medical-research-investment-takes-off-for-fisk-johnson/article_09bdfb27-1093-5b2b-8f22-60090a6899f5.html)

Money, Stewart. "Competition and the Future of the EELV program." The Space Review. December 12, 2011. (https://www.thespacereview.com/article/1990/1)

Money, Stewart. "Competition and the Future of the EELV program (Part 2)." The Space Review. December 12, 2011. (https://www.thespacereview.com/article/2042/2)

"Reusable Launch Vehicle Program Fact Sheet." NASA. September 1997. (https://www.hq.nasa.gov/office/pao/History/x-33/rlv_facts.htm)

Bergin, Chris. "X-33/VentureStar–What Really Happened." NASASpaceFlight.com. January 4, 2006. (https://www.nasaspaceflight.com/2006/01/x-33venturestar-what-really-happened/)

Marshall Space Flight Center Press Release. "Small Companies to Study Potential Use of Emerging Launch Systems for Alternative Access to Space Station." SpaceRef. August 24, 2000. (http://www.spaceref.com/news/viewpr.html?pid=2467)

NASA. "Launch Services Program: Earth's Bridge to Space" Brochure. 2012. (https://www.nasa.gov/sites/default/files/files/LSP_Brochure_508.pdf)

Dick, Steven J. and Roger D. Launius. *Critical Issues in the History of Spaceflight.* NASA Publication SP-2006-4702. Government Printing Office. 2006.

3. MODERN MYTHS

NASA Historical Data Book: Volume IV NASA Resources 1969-1978, SP-4012. (https://history.nasa.gov/SP-4012/vol4/contents.html)

Day, Dwayne A., PhD. "A Historic Meeting on Spaceflight... Background and Analysis." NASA. (https://history.nasa.gov/JFK-Webbconv/pages/backgnd.html)

John F. Kennedy Presidential Library and Museum Press Release. "JFK Library Releases Recording of President Kennedy Discussing Race to the Moon." May 25, 2011. (https://www.jfklibrary.org/about-us/news-and-press/press-releases/jfk-library-releases-recording-of-president-kennedy-discussing-race-to-the-moon)

Kennedy, President John F. "Address Before the 18th General Assembly of the United Nations, September 20, 1963." John F. Kennedy Presidential Library and Museum. jfklibrary.org. (https://www.jfklibrary.org/archives/other-resources/john-f-kennedy-speeches/united-nations-19630920)

"The Moon Decision." Apollo to the Moon Exhibition, Online Text. Smithsonian National Air and Space Museum. (https://airandspace.si.edu/exhibitions/apollo-to-the-moon/online/racing-to-space/moon-decision.cfm)

Moonrise Podcast. Hosted by Lillian Cunningham. *The Washington Post.* 2019. (https://www.washingtonpost.com/graphics/2019/national/podcasts/moonrise-the-origins-of-apollo-11-mission/)

Straus, Lawrence Guy (editor). "Projecting Favorable Perceptions of Space." *Journal of Anthropological Research. The University of Chicago Press Journals.* (https://www.journals.uchicago.edu/journals/jar/pr/201020)

"Margaret Mead: Human Nature and the Power of Culture" Exhibition, Online Text. Library of Congress. (https://www.loc.gov/exhibits/mead/oneworld-learn.html)

Dickson, Paul. "A Blow to the Nation." NOVA. (https://www.pbs.org/wgbh/nova/sputnik/nation.html)

"Declassified CIA Papers Show U.S. Aware in Advance of Sputnik Possibilities." RadioFreeEurope, RadioLiberty. October 5, 2017. (https://www.rferl.org/a/sputnik-cia-papers-anniversary/28774855.html)

Fortin, Jacey. "When Soviets Launched Sputnik, C.I.A. Was Not Surprised." *The New York Times.* October 6, 2017. (https://www.nytimes.com/2017/10/06/science/sputnik-launch-cia.html)

"Inquiry into Satellite and Missile Programs: Hearing Before the Preparedness Investigating Subcommittee of the Committee on Armed Services." United States Senate, Eighty-fifth Congress, first and second sessions. Government Printing Office. 1958.

Fishman, Charles. "How the First U.S. Satellite Launch Became Something of an International Joke." *Fast Company.* June 4, 2019. (https://www.fastcompany.com/90358292/how-the-first-u-s-satellite-launch-became-something-of-an-international-joke)

"Editorial Comment on the Nation's Failure to Launch a Test Satellite." *The New York Times.* December 8, 1957. (https://timesmachine.nytimes.com/timesmachine/1957/12/08/113410150.html?pageNumber=36)

Cordiner, Ralph J. "Competitive Private Enterprise in Space." In *Peacetime Uses of Outer Space*, edited by Simon Ramo. McGraw-Hill Book Company, Inc. 1961. (https://rjacobson.files.wordpress.com/2011/02/cordiner-article-1961.pdf)

Transcript of President Dwight D. Eisenhower's Farewell Address. 1961. (https://www.ourdocuments.gov/doc.php?flash=false&doc=90&page=transcript)

Brown, Archie. *The Human Factor: Gorbachev, Reagan, and Thatcher and the End of the Cold War*. Oxford University Press. 2020.

Oreskes, Naomi and Erick M. Conway. *Merchants of Doubt: How a Handful of Scientists Obscured the Truth on Issues from Tobacco Smoke to Global Warming*. Bloomsbury Press. 2010.

Tedeschi, Diane. "How Much Did Wernher von Braun Know, and When Did He Know It?" *Air & Space Magazine*. January 1, 2008. (https://www.airspacemag.com/space/a-amp-s-interview-michael-j-neufeld-23236520/)

Lehrer, Tom. Lyrics to "Wernher von Braun." Video of song performance: youtube.com/watch?v=QEJ9HrZq7Ro

4. RISKY BUSINESS

Wilson, Jim (editor). "Shuttle-Mir." NASA. (https://www.nasa.gov/mission_pages/shuttle-mir/)

Smith, Marcia S. "Space Stations." Congressional Research Brief for Congress. November 17, 2005. (https://sgp.fas.org/crs/space/IB93017.pdf)

Boudette, Neal E. "Space Buffs Attempt to Make Their Mir Tourist Venture Fly." *The Wall Street Journal*. June 16, 2000. (https://www.wsj.com/articles/SB961108659834371139)

Foust, Jeff. "AstroMom and Basstronaut, revisited." The Space Review. November 19, 2007. (https://www.thespacereview.com/article/1003/1)

Potter, Ned. "Boy Band, Astro Mom Battle for Space Tourist Spot." ABC News. January 7, 2006. (https://abcnews.go.com/WNT/story?id=130417&page=1)

Columbia Accident Investigation Board. "Report of Columbia Accident Investigation Board, Volume I." NASA. August 6, 2003. (https://www.nasa.gov/columbia/home/CAIB_Vol1.html)

Sunseri, Gina. "Columbia Shuttle Crew Not Told of Possible Problem With Reentry." ABC News. January 31, 2013. (https://abcnews.go.com/Technology/columbia-shuttle-crew-told-problem-reentry/story?id=18366185)

Pianin, Eric and Kathy Sawyer. "Denial of Shuttle Image Requests Questioned." *The Washington Post*. April 9, 2003. (https://www.washingtonpost.com/archive/politics/2003/04/09/denial-of-shuttle-image-requests-questioned/80957e7c-92f1-48ae-8272-0dcfbcb57b9d/)

Rensberger, Boyce and Kathy Sawyer. "Challenger Disaster Blamed on O-Rings, Pressure to Launch." *The Washington Post*. June 10, 1986. (https://www.washingtonpost.com/archive/politics/1986/06/10/challenger-disaster-blamed-on-o-rings-pressure-to-launch/6b331ca1-f544-4147-8e4e-941b7a7e47ae/)

Kay, W.D. "Democracy and Super Technologies: The Politics of the Space Shuttle and Space Station *Freedom*." *Science, Technology, and Human Values,* Volume 19, No. 2. Sage Publications Inc. 1994.

Airlines for America. "Safety Record of U.S. Air Carriers." Data & Statistics. November 11, 2021. (https://www.airlines.org/dataset/safety-record-of-u-s-air-carriers/)

"Annual Passengers on All U.S. Scheduled Airline Flights (Domestic & International) and Foreign Airline Flights to and from the United States, 2003-2018." Bureau of Transportation Statistics. (https://www.bts.dot.gov/annual-passengers-all-us-scheduled-airline-flights-domestic-international-and-foreign-airline)

"Active Duty Military Deaths by Year and Manner, 1980–2010 (As of November 2011)." Defense Casualty Analysis System. Defense Data Manpower Center. (https://dcas.dmdc.osd.mil/dcas/pages/report_by_year_manner.xhtml)

Ritchie, Erika I. "More US Service Members Die Training Than at War. Can the Pentagon Change That?" Task & Purpose. May 13, 2018. (https://taskandpurpose.com/analysis/military-training-accidents-aviation/)

U.S. House of Representatives Committee on Science, Space, & Technology Press Release. "GAO Report Finds Failure of Oversight by NASA IG." January 9, 2009. (https://science.house.gov/news/press-releases/gao-report-finds-failure-of-oversight-by-nasa-ig)

Brinkerhoff, Noel. "Failed NASA Inspector General Finally Resigns." AllGov. April 4, 2009. (http://www.allgov.com/news/appointments-and-resignations/failed-nasa-inspector-general-finally-resigns?news=838529)

5. LOOKING UNDER THE HOOD

SpaceNews editor. "Clinton Team Stresses Balance for NASA During Fundraising Event." *SpaceNews*. June 29, 2004. (https://spacenews.com/clinton-team-stresses-balance-nasa-during-fundraising-event/)

Foust, Jeff. "The So-so Space Debate." The Space Review. June 2, 2008. (https://www.thespacereview.com/article/1142/1)

United States Government Accountability Office. "NASA: Agency Has Taken Steps Toward Making Sound Investment Decisions for Ares I But Still Faces Challenging Knowledge Gaps." October 2007. (https://www.gao.gov/assets/gao-08-51.pdf)

Committee to Review NASA's Exploration Technology Development Program; Aeronautics and Space Engineering Board; Diversion on Engineering and Physical Sciences; National Research Council of the National Academies. *A Constrained Space Exploration Technology Program: A Review of NASA's Exploration Technology Development Program*. National Academies Press. 2008.

Jones, Richard M. "GAO Questions NASA's Management of Constellation Program." FYI: Science Policy News from AIP. American Institute of Physics. October 16, 2009. (https://www.aip.org/fyi/2009/gao-questions-nasa's-management-constellation-program)

NASA. "Challenges in Completing and Sustaining the International Space Station." Report Number GAO-08-581T. April 24, 2008. (https://www.govinfo.gov/content/pkg/GAOREPORTS-GAO-08-581T/html/GAOREPORTS-GAO-08-581T.htm)

NASA. "NASA: Constellation Program Cost and Schedule Will Remain Uncertain Until a Sound Business Case Is Established." Report Number GAO-09-844. August 2009. (https://www.gao.gov/assets/a294329.html)

Davis, Jason. "'Apollo on Steroids': The Rise and Fall of NASA's Constellation Moon Program." The Planetary Society. August 1, 2016. (https://www.planetary.org/articles/20160801-horizon-goal-part-2)

House Committee on Science, Space, and Technology: Status Report. "Testimony by Norman Augustine Hearing on 'Options and Issues for NASA's Human Space Flight Program: Report of the Review of U.S. Human Space Flight Program.'" SpaceRef. September 15, 2009. (http://www.spaceref.com/news/viewsr.html?pid=32379)

Rutherford, Emelie. "Dim Outlook for Constellation Program in Augustine Panel's Report." Defense Daily. August 13, 2009. (https://www.defensedaily.com/dim-outlook-for-constellation-program-in-augustine-panels-report/congress/)

Messier, Doug. "A Look at Cost Overuns and Schedule Delays in Major Space Programs." Parabolic Arc. May 4, 2011. (http://www.parabolicarc.com/2011/05/04/cost-overruns/)

Klamper, Amy. "Ares 1 Advocates Take on Commercialization Proponents." *SpaceNews*. November 9, 2009. (https://spacenews.com/ares-1-advocates-take-commercialization-proponents/)

"NASA Chief Questions Urgency of Global Warming." *Morning Edition*. NPR. May 31, 2007. (https://www.npr.org/templates/story/story.php?storyId=10571499)

Block, Robert and Mark K. Matthews and *Sentinel* Staff Writers. "NASA Chief Griffin Bucks Obama's Transition Team." *Orlando Sentinel*. December 11, 2008. (https://www.orlandosentinel.com/news/os-xpm-2008-12-11-nasa11-story.html)

Stover, Dawn. "Obama Clashes with NASA Moon Program." *Popular Science*. December 12, 2008. (https://www.popsci.com/military-aviation-amp-space/article/2008-12/chicago-we-have-problem/)

Kluger, Jeffrey. "Does Obama Want to Ground NASA's Next Moon Mission?" *Time*. December 11, 2008. (http://content.time.com/time/nation/article/0,8599,1866045,00.html)

Benen, Steve. "Transition Trouble at NASA." *Washington Monthly*. December 11, 2008. (https://washingtonmonthly.com/2008/12/11/transition-trouble-at-nasa/)

Borenstein, Seth. "NASA Chief's Wife: Don't Fire My Husband." NBC News. December 31, 2008. (https://www.nbcnews.com/id/wbna28451925)

Cowing, Keith. "Major General Jonathan Scott Gration Emerges as Possible Obama Choice for NASA Administrator." SpaceRef. January 13, 2009. (http://www.spaceref.com/news/viewnews.html?id=1316)

Iannotta, Becky. "Key U.S. Senator Cautions Obama on NASA Pick." Space.com. January 14, 2009. (https://www.space.com/6313-key-senator-cautions-obama-nasa-pick.html)

U.S. Department of Transportation. "Worldwide Commercial Space Launches." Bureau of Transportation Statistics. (https://www.bts.gov/content/worldwide-commercial-space-launches)

National Aeronautics and Space Administration. "The Vision for Space Exploration." February 2004. (https://www.nasa.gov/pdf/55583main_vision_space_exploration2.pdf)

"The Air Mail Act of 1925 (Kelly Act)." Aviation Online Magazine. (http://avstop.com/history/needregulations/act1925.htm)

Berger, Brian. "SpaceX, Rocketplane Kistler Win NASA COTS Competition." Space.com. August 18, 2006. (https://www.space.com/2768-spacex-rocketplane-kistler-win-nasa-cots-competition.html)

Whitesides, Loretta Hildago. "NASA Terminates COTS Funds for Rocketplane Kistler." *Wired.* September 18, 2007. (https://www.wired.com/2007/09/nasa-terminates/)

SpaceX Press Release. "SpaceX: Support NASA Exploration and COTS Capability D." SpaceRef. February 11, 2009. (http://www.spaceref.com/news/viewpr.html?pid=27552)

"Commercial Crew & Cargo." NASA. (https://www.nasa.gov/offices/c3po/about/c3po.html)

Sargent Jr., John F., Coordinator. "Federal Research and Development Funding: FY2010." CRS Report for Congress. Congressional Research Service. January 12, 2010. (https://sgp.fas.org/crs/misc/R40710.pdf)

Ionnotta, Becky. "Multiple Options Available with Stimulus Money for NASA." *SpaceNews.* March 6, 2009. (https://spacenews.com/multiple-options-available-stimulus-money-nasa/)

Sausser, Brittany. "NASA Uses Stimulus Funding for Commercial Crew Concepts." *MIT Technology Review.* August 6, 2009. (https://www.technologyreview.com/2009/08/06/211114/nasa-uses-stimulus-funding-for-commercial-crew-concepts/)

Bolden, Charles F., Interviewed by Sandra Johnson. "NASA Johnson Space Center Oral History Project, Edited Oral History Transcript." January 15, 2004. (https://historycollection.jsc.nasa.gov/JSCHistoryPortal/history/oral_histories/BoldenCF/BoldenCF_1-15-04.htm)

Review of U.S. Human Spaceflight Plans Committee. "Seeking a Human Spaceflight Program Worth of a Great Nation." October 2009. NASA. (https://www.nasa.gov/pdf/396093main_HSF_Cmte_FinalReport.pdf)

Bettex, Morgan. "Reporter's Notebook: Where Do We Go from Here?" MIT News. December 16, 2009. (https://news.mit.edu/2009/notebook-augustine-1216)

NASA Fiscal Year 2011 Budget Estimates. NASA. (https://www.nasa.gov/pdf/420990main_FY_201_%20Budget_Overview_1_Feb_2010.pdf)

Malik, Tariq. "Obama Budget Scraps NASA Moon Plan for '21st Century Space Program.'" Space.com. February 1, 2010. (https://www.space.com/7849-obama-budget-scraps-nasa-moon-plan-21st-century-space-program.html)

Sacks, Ethan. "Lost in Space: President Obama's Proposed Budget Scraps NASA's Planned Manned Missions to the Moon." *New York Daily News*. February 1, 2010. (https://www.nydailynews.com/news/politics/lost-space-president-obama-proposed-budget-scraps-nasa-planned-manned-missions-moon-article-1.196064)

Jones, Richard M. "Senator Nelson on NASA's FY 2011 Budget Request." FYI: Science Policy News from AIP. American Institute of Physics. February 18, 2010. (https://www.aip.org/fyi/2010/senator-nelson-nasa's-fy-2011-budget-request)

Maliq, Tarik. "NASA Grieves Over Canceled Program." NBC News. February 2, 2010. (https://www.nbcnews.com/id/wbna35209628)

"Florida Congressional Delegation Letter to President Obama Regarding NASA FY 2011 Budget." US House of Representatives. SpaceRef. March 4, 2010. (http://www.spaceref.com/news/viewsr.html?pid=33634)

Werner, Debra. "Senators Decry NASA's Change of Plans." *SpaceNews*. February 25, 2010. (https://spacenews.com/senators-decry-nasas-change-plans/)

Klamper, Amy. "Garver: Battle Over Obama Plan Imperils NASA Budget Growth." *SpaceNews*. March 5, 2010. (https://spacenews.com/garver-battle-over-obama-plan-imperils-nasa%E2%80%82budget-growth/)

Klamper, Amy. "NASA Prepares 'Plan B' for New Space Plan." Space.com. March 4, 2010. (https://www.space.com/8002-nasa-prepares-plan-space-plan.html)

Chang, Kenneth. "NASA Chief Denies Talk of Averting Obama Plan." *The New York Times*. March 4, 2010. (https://www.nytimes.com/2010/03/05/science/space/05nasa.html)

Coats, Michael L., Interviewed by Jennifer Ross-Nazzal. "NASA Johnson Space Center Oral History Project, Edited Oral History Transcript." August 5, 2015. (https://historycollection.jsc.nasa.gov/JSCHistoryPortal/history/oral_histories/CoatsML/CoatsML_8-5-15.htm)

Lambwright, W. Henry. "Reflections on Leadership and Its Politics: Charles Bolden, NASA Administrator, 2009–2017." *Public Administration Review*. Syracuse University. July/August 2017.

6. HEAVY LIFT

Yeomans, Donald K. "Why Study Asteroids." Solar System Dynamics. April 1998. (https://ssd.jpl.nasa.gov/?why_asteroids)

Klamper, Amy. "Obama's NASA Overhaul Encounters Continued Congressional Resistance." *SpaceNews*. April 23, 2010. (https://spacenews.com/obamas-nasa%E2%80%82overhaul-encounters-continued-congressional-resistance/)

CNN Wire Staff. "Obama Outlines New NASA Strategy for Deep Space Exploration." CNN Politics. April 15, 2010. (http://www.cnn.com/2010/POLITICS/04/15/obama.space/index.html)

The White House, Office of the Press Secretary. "Remarks by the President on Space Exploration in the 21st Century." John F. Kennedy Space Center, Merritt Island, Florida. April 15, 2010. (https://www.nasa.gov/news/media/trans/obama_ksc_trans.html)

Chang, Kenneth. "Obama Vows Renewed Space Program." *The New York Times*. April 15, 2010. (https://www.nytimes.com/2010/04/16/science/space/16nasa.html)

Malik, Tariq. "Obama Aims to Send Astronauts to an Asteroid, Then to Mars." Space.com. April 15, 2010. (https://www.space.com/8222-obama-aims-send-astronauts-asteroid-mars.html)

President Barack Obama tours SpaceX with CEO Elon Musk. Photograph and Caption. The White House Archives. (https://obamawhitehouse.archives.gov/photos-and-video/photos/president-barack-obama-tours-spacex-with-ceo-elon-musk)

Matthews, Mark K. and Robert Block and *Orlando Sentinel*. "Obama Unveils NASA 'Vision' in Kennedy Space Center Speech." *Orlando Sentinel*. April 16, 2010. (https://www.orlandosentinel.com/news/os-xpm-2010-04-16-os-obama-speech-kennedy-space-center-20100415-story.html)

Moskowitz, Clara. "NASA Should Use Private Spaceships, Say Astronauts." *The Christian Science Monitor*. July 16, 2010. (https://www.csmonitor.com/Science/2010/0716/NASA-should-use-private-spaceships-say-astronauts)

Public Law 111–267. 111th Congress. "National Aeronautics and Space Administration Authorization Act of 2010." October 11, 2010. (https://www.congress.gov/111/plaws/publ267/PLAW-111publ267.pdf)

The National Aeronautics and Space Administration. "National Aeronautics and Space Act of 1958, As Amended." August 25, 2008. (https://history.nasa.gov/spaceact-legishistory.pdf)

Foust, Jeff. "Utah Members Concerned NASA 'Circumventing the Law' on Heavy Lift." Space Politics. November 19, 2010. (http://www.spacepolitics.com/2010/11/19/utah-members-concerned-nasa-circumventing-the-law-on-heavy-lift/)

Foust, Jeff. "Senate Carries Out Its Subpoena Threat." Space Politics. July 28, 2011. (http://www.spacepolitics.com/2011/07/28/senate-carries-out-its-subpoena-threat/)

Klotz, Irene. "NASA Sending Retired Space Shuttles to US Museums." Reuters. April 13, 2011. (https://www.reuters.com/article/uk-space-shuttles/nasa-sending-retired-space-shuttles-to-us-museums-idUSLNE73C02H20110413)

NASA Content Administrator. "NASA Transfers Enterprise Title to Intrepid Sea, Air & Space Museum in New York City." December 11, 2011. (https://www.nasa.gov/mission_pages/transition/placement/enterprise_transfer.html)

"Retired Space Shuttle Makes Final Voyage." VOA News. April 16, 2012. (https://www.voanews.com/a/retired-space-shuttle-makes-final-voyage-to-washington-area-museum-147761985/180532.html)

NASA Content Administrator. "Endeavour's Final Flight Ends." September 21, 2012. (https://www.nasa.gov/multimedia/imagegallery/endeavour_garver.html)

Achenbach, Joel. "Final NASA Shuttle Mission Clouded by Rancor." *The Washington Post*. July 2, 2011. (https://www.washingtonpost.com/national/health-science/us-space-program-approaches-end-of-an-era-what-next/2011/06/29/AGeBAWtH_story.html)

Booze, Allen, and Hamilton. Executive Summary of Final Report. "Independent Cost Assessment of the Space Launch System, Multi-Purpose Crew Vehicle and 21st

Century Ground Systems Programs." (https://www.nasa.gov/pdf/581582main_BAH_Executive_Summary.pdf)

US Senate Live Webcast. "Space Launch System Design Announced." September 14, 2011. (https://www.youtube.com/watch?v=TVp6uKfR5qE)

Leone, Dan. "Obama Administration Accused of Sabotaging Space Launch System." Space.com. September 12, 2011. (https://www.space.com/12916-obama-nasa-space-launch-system-budget.html)

Luscombe, Richard. "Nasa Shows Off 'Most Powerful Space Rocket in History.'" The Guardian. September 14, 2011. (https://www.theguardian.com/science/2011/sep/14/nasa-space-launch-system)

Space.com Staff. "Voices: Industry & Analysts Weigh In on NASA's New Rocket." Space.com. September 15, 2011. (https://www.space.com/12959-nasa-space-launch-system-rocket-reactions.html)

Foust, Jeff. "A Monster Rocket, or Just a Monster?" The Space Review. September 19, 2011. (https://www.thespacereview.com/article/1932/1)

NASA Press Release. "NASA to Brief Industry on Space Launch System Procurement." September 23, 2011. (https://www.nasa.gov/home/hqnews/2011/sep/HQ_M11-204_MSFC_Indust_Day.html)

Plait, Phil. "Why NASA Still Can't Put Humans in Space: Congress Is Starving It of Needed Funds." Slate. August 24, 2015. (https://slate.com/technology/2015/08/congress-and-nasa-commercial-crew-program-is-underfunded.html)

Obama, Barack. *A Promised Land*. Crown. 2020.

7. DARK MATTER

Kelly, Emre. "GAO Takes Aim at NASA's James Webb Space Telescope, Notes Delays and Cost Overruns." *Florida Today*. January 31, 2020. (https://www.floridatoday.com/story/tech/science/space/2020/01/31/gao-takes-aim-nasa-james-webb-space-telescope-delays-cost-overruns/4624433002/)

Moskowitz, Clara. "NASA's Next Mars Rover Still Faces Big Challenges, Audit Reveals." Space.com. June 8, 2011. (https://www.space.com/11903-mars-rover-curiosity-budget-delay-report.html)

Weiler, Edward J., Interviewed by Sandra Johnson. "NASA Science Mission Directorate Oral History Project, Edited Oral History Transcript." April 4, 2017. (https://historycollection.jsc.nasa.gov/JSCHistoryPortal/history/oral_histories/NASA_HQ/SMD/WeilerEJ/WeilerEJ_4-4-17.htm)

United States Government Accountability Office. "NASA Needs to Better Assess Contract Termination Liability Risks and Ensure Consistency in Its Practices." Report. July 12, 2011. (https://www.gao.gov/products/gao-11-609r)

Klamper, Amy. "Obama's NASA Overhaul Encounters Continued Congressional Resistance." *SpaceNews*. April 23, 2010. (https://spacenews.com/obamas-nasa%E2%80%82overhaul-encounters-continued-congressional-resistance/)

NASA Advisory Council Recommendation. Industrial Base 2011-02-04 (EC-03). Attached: June 2011 NASA report to Congress, "Effects of the Transition to the

Space Launch System on the Solid and Liquid Rocket Motor Industrial Bases." (https://www.nasa.gov/sites/default/files/atoms/files/may2011_industrialbase.pdf)

Lambwright, W. Henry. "Reflections on Leadership and Its Politics: Charles Bolden, NASA Administrator, 2009–2017." *Public Administration Review*. Syracuse University. July/August 2017.

"Charles Bolden, the NASA Administrator and Astronaut in Conversation with Al Jazeera's Imran Garda." Talk to Al Jazeera. Al Jazeera. July 1, 2010. (https://www.aljazeera.com/program/talk-to-al-jazeera/2010/7/1/charles-bolden)

Moskowitz, Clara. "NASA Chief Says Agency's Goal Is Muslim Outreach, Forgets to Mention Space." *The Christian Science Monitor*. July 14, 2010. (https://www.csmonitor.com/Science/2010/0714/NASA-chief-says-agency-s-goal-is-Muslim-outreach-forgets-to-mention-space)

Reuters Staff. "White House Corrects NASA Chief on Muslim Comment." Reuters. July 12, 2010. (https://www.reuters.com/article/us-obama-nasa/white-house-corrects-nasa-chief-on-muslim-comment-idUSTRE66B6MQ20100712)

Foust, Jeff. "Suborbital Research Enters a Time of Transition." The Space Review. June 10, 2013. (https://www.thespacereview.com/article/2311/1)

NASA. "OMEGA Project 2009–2012." (https://www.nasa.gov/centers/ames/research/OMEGA/index.html)

Foust, Jeff. "Former NASA Administrator Reprimanded for Use of Agency Personnel After Departure." *SpaceNews*. June 12, 2020. (https://spacenews.com/former-nasa-administrator-reprimanded-for-use-of-agency-personnel-after-departure/)

8. RISE OF THE ROCKETEERS

"Card, Club, Pan Am 'First Moon Flights.'" Pan Am's Club Card for "First Moon Flights." Number 1043, issued by the airline to Jeffrey Gates. Smithsonian National Air and Space Museum. (https://airandspace.si.edu/collection-objects/card-club-pan-am-first-moon-flights/nasm_A20180010000)

Borcover, Alfred and Travel Editor. "161 Hopefuls Put Up $5,000 Each to Experience an Out-of-this-World Trip." *Chicago Tribune*. January 12, 1986. (https://www.chicagotribune.com/news/ct-xpm-1986-01-12-8601040135-story.html)

McCray, W. Patrick. *The Visioneers: How a Group of Elite Scientists Pursued Space Colonies, Nanotechnologies, and a Limitless Future*. Princeton University Press. 2017. (http://assets.press.princeton.edu/chapters/i9822.pdf)

Fowler, Glenn. "George Koopman Dies in Wreck; Technologist for Space Was 44." *The New York Times*. July 21, 1989. (https://www.nytimes.com/1989/07/21/obituaries/george-koopman-dies-in-wreck-technologist-for-space-was-44.html)

Boyle, Alan. "Space Racers Unite in Federation." NBC News. February 8, 2005. (https://www.nbcnews.com/id/wbna6936543)

Boyle, Alan. "Private-Spaceflight Bill Signed into Law." NBC News. December 8, 2004. (https://www.nbcnews.com/id/wbna6682611)

"Rocket Man." *Forbes*. April 17, 2000. (https://www.forbes.com/forbes/2000/0417/6509398a.html?sh=3fe63e906d4b)

Clark, Stephen. "Beal Aerospace Ceases Work to Build Commercial Rocket." Spaceflight Now. October 24, 2000. (https://spaceflightnow.com/news/n0010/24beal/)

Chang, Kenneth. "For Space Station, a Pod That Folds Like a Shirt and Inflates Like a Balloon." *The New York Times*. January 16, 2013. (https://www.nytimes.com/2013/01/17/science/space/for-nasa-bigelow-aerospaces-balloonlike-module-is-innovative-and-a-bargain-too.html)

Foust, Jeff. "Bigelow Aerospace Lays Off Entire Workforce." *SpaceNews*. March 23, 2020. (https://spacenews.com/bigelow-aerospace-lays-off-entire-workforce/)

Foust, Jeff. "Stratolaunch Founder Paul Allen Dies." *SpaceNews*. October 15, 2018. (https://spacenews.com/stratolaunch-founder-paul-allen-dies/)

Abdollah, Tami and Stuart Silverstein. "Test Site Explosion Kills Three." *Los Angeles Times*. July 27, 2007. (https://www.latimes.com/archives/la-xpm-2007-jul-27-me-explode27-story.html)

Malik, Tariq. "Deadly SpaceShipTwo Crash Caused by Co-Pilot Error: NTSB." Space.com. July 28, 2015. (https://www.space.com/30073-virgin-galactic-spaceshiptwo-crash-pilot-error.html)

Malik, Tariq. "Virgin Galactic Goes Public on New York Stock Exchange After Completing Merger." Space.com. October 28, 2019. (https://www.space.com/virgin-galactic-goes-public-nyse-stock-exchange.html)

Weitering, Hanneke. "Blue Moon: Here's How Blue Origin's New Lunar Lander Works." Space.com. May 10, 2019. (https://www.space.com/blue-origin-blue-moon-lander-explained.html)

Thomas, Candrea. "Blue Origin Tests Rocket Engine Thrust Chamber." Commercial Space Transportation. NASA. October 15, 2012. (https://www.nasa.gov/exploration/commercial/crew/blue-origin-be3.html)

Berger, Brian. "SpaceX, Rocketplane Kistler Win NASA COTS Competition." Space.com. August 18, 2006. (https://www.space.com/2768-spacex-rocketplane-kistler-win-nasa-cots-competition.html)

Chang, Kenneth. "First Private Craft Docks with Space Station." *The New York Times*. May 25, 2012. (https://www.nytimes.com/2012/05/26/science/space/space-x-capsule-docks-at-space-station.html)

Killian, Mike. "Government Requests Court Dismiss SpaceX Lawsuit Over Air Force's 36-Rocket Block-Buy Deal With ULA." AmericaSpace. 2014. (https://www.americaspace.com/2014/07/03/government-requests-court-dismiss-spacex-lawsuit-over-air-forces-36-rocket-block-buy-deal-with-ula/)

Gruss, Mike. "SpaceX, Air Force Settle Lawsuit over ULA Blockbuy." *SpaceNews*. January 23, 2015. (https://spacenews.com/spacex-air-force-reach-agreement/)

Berger, Eric. "This Is Probably Why Blue Origin Keeps Protesting NASA's Lunar Lander Award." Ars Technica. August 11, 2021. (https://arstechnica.com/science/2021/08/this-is-probably-why-blue-origin-keeps-protesting-nasas-lunar-lander-award/)

"TODAY: SpaceX to Make First Launch Attempt for COTS Demo 1." SpaceRef. December 8, 2010. (http://www.spaceref.com/news/viewpr.html?pid=32213)

Junod, Tom. "Elon Musk: Triumph of His Will." *Esquire*. November 15, 2012. (https://www.esquire.com/news-politics/a16681/elon-musk-interview-1212/)

Sauser, Brittany. "SpaceX Sets Launch for Heavy-Lift Rocket." *Technology Review*. April 5, 2011. (https://www.technologyreview.com/2011/04/05/195936/spacex-sets-launch-date-for-heavy-lift-rocket/)

Garver, Lori. "SpaceX Could Save NASA and the Future of Space Exploration." Op-ed. *The Hill*. February 8, 2018. (https://thehill.com/opinion/technology/372994-spacex-could-save-nasa-and-the-future-of-space-exploration)

Mwaniki, Andrew. "Countries with the Most Commercial Space Launches." World Atlas. May 16, 2018. (https://www.worldatlas.com/articles/countries-with-the-most-commercial-space-launches.html)

9. IT'S NOT JUST ROCKET SCIENCE

"The Dawn of the Space Shuttle." Richard Nixon Foundation. January 5, 2017. (https://www.nixonfoundation.org/2017/01/dawn-space-shuttle/)

Noë, Alva. "Soaking Up Wisdom from Neil DeGrasse Tyson." NPR. January 22, 2016. (https://www.npr.org/sections/13.7/2016/01/22/463855900/soaking-up-wisdom-from-neil-degrasse-tyson)

Tyson, Neil deGrasse. "Neil deGrasse Tyson: The 3 Fears That Drive Us to Accomplish Extraordinary Things." Big Think. July 19, 2013. (https://youtu.be/0CJ8g8w1huc)

NASA. "Our Missions and Values." (https://www.nasa.gov/careers/our-mission-and-values)

Coats, Michael L., Interviewed by Jennifer Ross-Nazzal. "NASA Johnson Space Center Oral History Project, Edited Oral History Transcript." August 5, 2015. (https://historycollection.jsc.nasa.gov/JSCHistoryPortal/history/oral_histories/CoatsML/CoatsML_8-5-15.htm)

"Model, X-33 VentureStar Reusable Launch Vehicle." Transferred from NASA Langley Research Center. Smithsonian National Air & Space Museum. (https://airandspace.si.edu/collection-objects/model-x-33-venturestar-reusable-launch-vehicle/nasm_A20060581000)

Luypaert, Joris. "The Man Who Killed The X-33 Venturestar." One Stage to Space. June 23, 2018. (https://onestagetospace.com/2018/06/23/the-man-that-killed-the-x-33-venturestar/)

Oliva, Leandro. "Goodnight Moon: Michael Griffin on the future of NASA." Ars Technica. April 1, 2010. (https://arstechnica.com/science/2010/04/goodnight-moon-michael-griffin-on-the-future-of-nasa/)

Lambwright, W. Henry. "Reflections on Leadership and Its Politics: Charles Bolden, NASA Administrator, 2009–2017." *Public Administration Review*. Syracuse University. July/August 2017.

Ferguson, Sarah. "Launching Starship: Inside Elon Musk's Plan to Perfect the Rocket to Take Humanity to Mars." *Foreign Correspondent*. ABC (Australian Broadcasting Corporation) News. September 29, 2021. (https://www.abc.net.au/news/2021-09-30/elon-musk-starship-to-get-back-to-the-moon-and-on-to-mars/100498076)

Ferguson, Sarah. "Destination Mars." *Foreign Correspondent*. ABC (Australian Broadcasting Corporation) News. September 30, 2021. Updated November 1, 2021. (https://www.abc.net.au/foreign/destination-mars/13565384)

Statement of VADM Joseph W. Dyer, USN (Retired) Chairman National Aeronautics and Space Administration's Aerospace Safety Advisory Panel before the Committee on Science, Space, and Technology Subcommittee on Space and Aeronautics. U.S. House of Representatives. September 14, 2012. (https://www.hq.nasa.gov/legislative/hearings/2012%20hearings/9-14-2012%20DYER.pdf)

Regan, Rebecca. "NASA's Commercial Crew Program Refines Its Course." Commercial Space Transportation. NASA. December 21, 2011. (https://www.nasa.gov/exploration/commercial/crew/CCP_strategy.html)

Plait, Phil. "BREAKING: After Initial Problems, SpaceX Dragon Now Looking Good On Orbit." Slate. March 1, 2013. (https://slate.com/technology/2013/03/spacex-dragon-initial-problems-with-thrusters-now-under-control-mission-to-proceed-soon.html)

Gustetic, Jennifer L., Victoria Friedensen, Jason L. Kessler, Shanessa Jackson, and James Parr. "NASA's Asteroid Grand Challenge: Strategy, Results, and Lessons Learned." *Space Policy*. Volumes 44–45. August 2018. (https://www.sciencedirect.com/science/article/pii/S0265964617300838)

NASA. "Asteroid Redirect Mission Crewed Mission (ARCM) Concept Study." Mission Formulation Review. (https://www.nasa.gov/sites/default/files/files/Asteroid-Crewed-Mission-Stich-TAGGED2.pdf)

NASA Content Administrator. "Asteroid Mission Targeted." April 29, 2013. (https://www.nasa.gov/centers/dryden/news/X-Press/dfrc_budget_2013.html)

Berger, Eric. "NASA's Asteroid Mission Isn't Dead—Yet." Ars Technica. February 10, 2016. (https://arstechnica.com/science/2016/02/nasas-asteroid-mission-isnt-deadyet/)

10. TURNING WRONGS INTO RIGHTS

SpaceNews Staff. "Bolden Urges Work Force to Back NASA's New Direction." *SpaceNews*. May 3, 2010. (https://spacenews.com/bolden-urges-work-force-back-nasas-new-direction/)

Straus, Mark. "Majority of Americans Believe It Is Essential That the U.S. Remain a Global Leader in Space." Pew Research Center. June 6, 2018. (https://www.pewresearch.org/science/2018/06/06/majority-of-americans-believe-it-is-essential-that-the-u-s-remain-a-global-leader-in-space/)

Johnson, Courtney. "How Americans See the Future of Space Exploration, 50 Years After the First Moon Landing." Pew Research Center. July 17, 2019. (https://www.pewresearch.org/fact-tank/2019/07/17/how-americans-see-the-future-of-space-exploration-50-years-after-the-first-moon-landing/)

Sabin, Sam. "Nearly Half the Public Wants the U.S. to Maintain Its Space Dominance. Appetite for Space Exploration Is a Different Story." Morning Consult. February 25, 2021. (https://morningconsult.com/2021/02/25/space-force-travel-exploration-poll/)

Chase, Patrick. "NASA, Space Exploration, and American Public Opinion." WestEastSpace. Medium.com. July 14, 2020. (https://medium.com/westeastspace/nasa-space-exploration-and-american-public-opinion-139cbc1c6cce)

Treat, Jason, Jay Bennett, and Christopher Turner. "How 'The Right Stuff' Has Changed." *National Geographic*. November 6, 2020. (https://www.nationalgeographic.com/science/graphics/charting-how-nasa-astronaut-demographics-have-changed-over-time)

Krishna, Swapna. "The Mercury 13: The women who could have been NASA's first female astronauts." Space.com. July 24, 2020. (https://www.space.com/mercury-13.html)

Sylvester, Roshanna. "John Glenn and the Sexism of the Early Space Program." *Smithsonian Magazine*. December 14, 2016. (https://www.smithsonianmag.com/history/even-though-i-am-girl-john-glenns-fan-mail-and-sexism-early-space-program-180961443/)

Teitel, Amy Shira. "NASA Once Made an Official Ruling on Women and Pantsuits." *Discover*. February 12, 2019. (https://www.discovermagazine.com/the-sciences/nasa-once-made-an-official-ruling-on-women-and-pantsuits)

Shetterly, Margot Lee. *Hidden Figures: The American Dream and the Untold Story of the Black Women Mathematicians Who Helped Win the Space Race*. William Morrow. 2016.

Holt, Nathalia. *Rise of the Rocket Girls: The Women Who Propelled Us, from Missiles to the Moon to Mars*. Little, Brown and Company. 2016.

Katsarou, Maria. "Women & the Leadership Labyrinth Howard vs Heidi." Leadership Psychology Institute. (https://www.leadershippsychologyinstitute.com/women-the-leadership-labyrinth-howard-vs-heidi/)

Ottens, Nick. "Sexism in *Star Trek*." Forgotten Trek. October 16, 2019. (https://forgottentrek.com/sexism-in-star-trek/)

Ulster, Laurie. "15 Really Terrible Moments for Women in *Star Trek*." *Screen Rant*. August 1, 2016. (https://screenrant.com/terrible-moments-for-women-in-star-trek/)

Woman in Motion. Internet Movie Database. imdb.com/title/tt4512946/. (https://www.imdb.com/title/tt4512946/)

Coats, Michael L., Interviewed by Jennifer Ross-Nazzal. "NASA Johnson Space Center Oral History Project, Edited Oral History Transcript." August 5, 2015. (https://historycollection.jsc.nasa.gov/JSCHistoryPortal/history/oral_histories/CoatsML/CoatsML_8-5-15.htm)

Ride, Dr. Sally K. "Leadership and America's Future in Space: A Report to the Administrator." NASA. August 1987. (https://history.nasa.gov/riderep/main.PDF)

Discussion with Alan Ladwig. August 23, 2021.

Grady, Denise. "American Woman Who Shattered Space Ceiling: Sally Ride, 1951–2012." *The New York Times*. July 23, 2012. (https://www.nytimes.com/2012/07/24/science/space/sally-ride-trailblazing-astronaut-dies-at-61.html)

Sherr, Lynn. *Sally Ride: America's First Woman in Space*. Simon & Schuster. June 3, 2014.

Sorkin, Amy Davidson. "The Astronaut Bride." *The New Yorker*. July 25, 2012. (https://www.newyorker.com/news/daily-comment/the-astronaut-bride)

Davenport, Christian, and Rachel Lerman. "Inside Blue Origin: Employees Say Toxic, Dysfunctional 'Bro Culture' Led to Mistrust, Low Morale, and Delays at Jeff Bezos's Space Venture." *The Washington Post*. October 11, 2021. (https://www.washingtonpost.com/technology/2021/10/11/blue-origin-jeff-bezos-delays-toxic-workplace/)

Kolhatkar, Sheelah. "The Tech Industry's Gender-Discrimination Problem." *The New Yorker*. November 13, 2017. (https://www.newyorker.com/magazine/2017/11/20/the-tech-industrys-gender-discrimination-problem)

Tayeb, Zahra, and Kevin Shalvey. "Former SpaceX Engineer Accuses Company of Racial Discrimination, Denying Its Claims that He Was Fired for Making Inappropriate Facial Expressions." *Business Insider*. November 14, 2021. (https://www.businessinsider.com/spacex-engineer-alleges-racial-discrimination-harassment-lawsuit-2021-11)

Ivey, Glen E. "Lori Garver: Not Every Hero at NASA Is an Astronaut." gleneivey.wordpress.com. March 24, 2010. (https://gleneivey.wordpress.com/2010/03/24/lori-garver-not-every-hero-at-nasa-is-an-astronaut/)

Mackay, Charles. "No Enemies." 1846.

Sinclair, Upton. *Anthology*. Murray & Gee. 1947.

11. UNLEASHING THE DRAGON

Berger, Brian. "Outgoing NASA Deputy Reflects on High-profile, Big-money Programs." *SpaceNews*. September 9, 2013. (https://spacenews.com/37126outgoing-nasa-deputy-reflects-on-high-profile-big-money-programs/)

Chang, Kenneth. "Scrutinizing SpaceX, NASA Overlooked Some Boeing Software Problems." *The New York Times*. July 7, 2020. (https://www.nytimes.com/2020/07/07/science/boeing-starliner-nasa.html)

Davenport, Christian. "No One Thought SpaceX Would Beat Boeing. Elon Musk Proved Them Wrong." *The Washington Post*. May 21, 2020. (https://www.washingtonpost.com/technology/2020/05/21/spacex-boeing-rivalry-launch/)

Gohd, Chelsea. "NASA's SpaceX Launch Is Not the Cure for Racial Injustice." Space.com. June 3, 2020. (https://www.space.com/spacex-launch-not-cure-for-racial-injustice.html)

Heron, Gil Scott. "Whitey on the Moon." Spoken word poem. 1970.

Email from Kiko Dontchev, SpaceX. July 26, 2018.

"The Crew Dragon Mission Is a Success for SpaceX and for NASA." *The Economist*. June 6, 2020. (https://www.economist.com/science-and-technology/2020/06/04/the-crew-dragon-mission-is-a-success-for-spacex-and-for-nasa)

Wall, Mike. "Trump Hails SpaceX's 1st Astronaut Launch Success for NASA." Space.com. May 30, 2020. (https://www.space.com/trump-hails-spacex-astronaut-launch-demo-2.html)

Foust, Jeff. "Current and Former NASA Leadership Share Credit for Commercial Crew." *SpaceNews*. May 26, 2020. (https://spacenews.com/current-and-former-nasa-leadership-share-credit-for-commercial-crew/)

Erwin, Sandra. "Biden's Defense Nominee Embraces View of Space As a Domain of War." *SpaceNews*. January 19, 2021. (https://spacenews.com/bidens-defense-nominee-embraces-view-of-space-as-a-domain-of-war/)

Broad, William J. "How Space Became the Next 'Great Power' Contest Between the U.S. and China." *The New York Times*. January 24, 2021. Updated May 6, 2021. (https://www.nytimes.com/2021/01/24/us/politics/trump-biden-pentagon-space-missiles-satellite.html)

Howell, Elizabeth. "Jeff Bezos' Blue Origin Throws Shade at Virgin Galactic Ahead of Richard Branson's Launch." Space.com. July 9, 2021. (https://www.space.com/blue-origin-throws-shade-at-virgin-galactic-ahead-of-launch)

Hussain, Noor Zainab. "Branson's Virgin Galactic to Sell Space Tickets Starting at $450,000." Reuters. August 5, 2021. (https://www.reuters.com/lifestyle/science/bransons-virgin-galactic-sell-space-flight-tickets-starting-450000-2021-08-05/)

McFall-Johnsen, Morgan. "Elon Musk Showed Up in Richard Branson's Kitchen at 3 a.m. to Wish Him Luck Flying to the Edge of Space." *Business Insider*. July 11, 2021. (https://www.businessinsider.com/elon-musk-visited-richard-branson-kitchen-early-launch-day-2021-7)

Neuman, Scott. "Jeff Bezos and Blue Origin Travel Deeper into Space Than Richard Branson." NPR. July 20, 2021. (https://www.npr.org/2021/07/20/1017945718/jeff-bezos-and-blue-origin-will-try-to-travel-deeper-into-space-than-richard-bra)

Chang, Kenneth. "SpaceX Inspiration4 Mission: Highlights From Day 2 in Orbit." *The New York Times*. September 17, 2021. Updated November 9, 2021. (https://www.nytimes.com/live/2021/09/17/science/spacex-inspiration4-tracker)

Wall, Mike. "SpaceX to Fly 3 More Private Astronaut Missions to Space Station for Axiom Space." Space.com. June 2, 2021. (https://www.space.com/spacex-axiom-deal-more-private-astronaut-missions)

Johnson Space Center, Status Report. "NASA Commercial LEO Destinations Announcement 80JSC021CLD FINAL." NASA. SpaceRef. July 12, 2021. (http://www.spaceref.com/news/viewsr.html?pid=54956)

Powell, Corey S. "Jeff Bezos Foresees a Trillion People Living in Millions of Space Colonies. Here's What He's Doing to Get the Ball Rolling." NBC News. May 15, 2019. (https://www.nbcnews.com/mach/science/jeff-bezos-foresees-trillion-people-living-millions-space-colonies-here-ncna1006036)

Mosher, Dave. "Elon Musk Says SpaceX Is on Track to Launch People to Mars Within 6 Years. Here's the Full Timeline of His Plans to Populate the Red Planet." *Business Insider*. November 2, 2018. (https://www.businessinsider.com/elon-musk-spacex-mars-plan-timeline-2018-10)

Papadopoulos, Anna. "The World's Richest People (Top Billionaires, 2021)." *CEOWorld Magazine*. November 4, 2021. (https://ceoworld.biz/2021/11/04/the-worlds-richest-people-2021/)

Silverman, Jacob. "The Billionaire Space Race Is a Tragically Wasteful Ego Contest." *New Republic*. July 9, 2021. (https://newrepublic.com/article/162928/richard-branson-jeff-bezos-space-blue-origin)

Lepore, Jill. "Elon Musk Is Building a Sci-Fi World, and the Rest of Us Are Trapped in It." Guest Essay, Opinion. *The New York Times*. November 4, 2021. (https://www.nytimes.com/2021/11/04/opinion/elon-musk-capitalism.html)

Deggans, Eric. "Elon Musk Takes An Awkward Turn As 'Saturday Night Live' Host." *All Things Considered*. NPR. May 9, 2021. (https://www.npr.org/2021/05/09/994620764/elon-musk-hosts-snl)

Swift, Taylor Alison, and Jack Antonoff. "The Man." Lyrics. AZLyrics.com. (https://www.azlyrics.com/lyrics/taylorswift/theman.html)

12. THE VALUE PROPOSITION

Noble, Alex. "Jon Stewart Sends Up Billionaire Space Race in Starry Promo for New Apple Show." The Wrap. July 20, 2021. (https://www.thewrap.com/jon-stewart-jeff-bezos-space-race-apple-tv-show-promo-video/)

Baxter, William E. "Introduction—Samuel P. Langley: Aviation Pioneer (Part 2)." Smithsonian Libraries. (https://www.sil.si.edu/ondisplay/langley/part_two.htm)

Office of Science and Technology Policy, The White House. "Statement on National Space Transportation Policy." August 5, 1994. (https://www.globalsecurity.org/space/library/policy/national/launchst.htm)

NASA Office of Inspector General, Office of Audits. "NASA's Management of the Artemis Missions." Report No. IG-22-003. NASA. November 15, 2021. (https://oig.nasa.gov/docs/IG-22-003.pdf)

Davenport, Christian. "NASA Watchdog Takes Aim at Boeing's SLS Rocket; It Says Backbone of Trump's Moon Mission Could Cost a Staggering $50 Billion." *The Washington Post*. March 11, 2020. (https://www.washingtonpost.com/technology/2020/03/10/nasa-boeing-trump-moon-cost/)

Berger, Eric. "NASA Has Begun a Study of the SLS Rocket's Affordability [Updated]." Ars Technica. March 15, 2021. (https://arstechnica.com/science/2021/03/nasa-has-begun-a-study-of-the-sls-rockets-affordability/)

Miller, Amanda. "NASA Faces Up to Huge Cost Overruns for Its SLS Heavy Lift Rocket." *Room: Space Journal of Asgardia*. March 11, 2020. (https://room.eu.com/news/nasa-faces-up-to-huge-cost-overruns-for-its-sls-heavy-lift-rocket)

Wall, Mike. "'Artemis Is Here:' Vice President Pence Stresses Importance of 2024 Moon Landing." Space.com. November 14, 2019. (https://www.space.com/nasa-artemis-moon-program-mike-pence.html)

Sheetz, Michael. "Trump Wants NASA to Go to Mars, Not the Moon Like He Declared Weeks Ago." CNBC. June 7, 2019. (https://www.cnbc.com/2019/06/07/trump-wants-nasa-to-go-to-mars-not-the-moon-like-he-declared-weeks-ago.html)

Grush, Loren. "Trump Repeatedly Asks NASA Administrator Why We Can't Go Straight to Mars." The Verge. July 19, 2019. (https://www.theverge.com/2019/7/19/20701061/president-trump-nasa-administrator-jim-bridenstine-artemis-mars-direct-moon-apollo-11)

Davenport, Christian. "Trump Pushed for a Moon Landing in 2024. It's Not Going to Happen." *The Washington Post*. January 13, 2021. (https://www.washingtonpost.com/technology/2021/01/13/trump-nasa-moon-2024/)

Foust, Jeff. "Changing NASA Requirements Caused Cost and Schedule Problems for Gateway." *SpaceNews*. November 12, 2020. (https://spacenews.com/changing-nasa-requirements-caused-cost-and-schedule-problems-for-gateway/)

NASA Office of Inspector General, Office of Audits. "NASA's Development of Next-Generation Spacesuits." Report No. IG-21-025. NASA. August 10, 2021. (https://oig.nasa.gov/docs/IG-21-025.pdf)

Mahoney, Erin. "NASA Prompts Companies for Artemis Lunar Terrain Vehicle Solutions." NASA. August 31, 2021. (https://www.nasa.gov/feature/nasa-prompts-companies-for-artemis-lunar-terrain-vehicle-solutions)

Foust, Jeff. "Just off a call . . ." @jeff_foust Tweet. Twitter. May 26, 2020. 2:44 pm. (https://twitter.com/jeff_foust/status/1265353156206231556)

Rupar, Aaron. "Jen Psaki's Space Force Comment and the Ensuing Controversy, Explained." Vox. February 5, 2021. (https://www.vox.com/2021/2/5/22268047/jen-psaki-space-force)

Foust, Jeff. "White House Endorses Artemis Program." *SpaceNews*. February 4, 2021. (https://spacenews.com/white-house-endorses-artemis-program/)

Berger, Eric. "White House Says Its Supports Artemis Program to Return to the Moon [Updated]." Ars Technica. February 4, 2021. (https://arstechnica.com/science/2021/02/senate-democrats-send-a-strong-signal-of-support-for-artemis-moon-program/)

Smith, Marcia. "Biden Administration 'Certainly' Supports Artemis Program." SpacePolicyOnline.com. February 4, 2021. (https://spacepolicyonline.com/news/biden-administration-certainly-supports-artemis-program/)

Garver, Lori. "New NASA Administrator Should Reject Its Patriarchal and Parochial Past." Op-ed. *Scientific American*. April 12, 2021. (https://www.scientificamerican.com/article/bill-nelson-isnt-the-best-choice-for-nasa-administrator/)

Axe, David. "NASA Veterans Baffled by Biden Pick of Bill Nelson to Lead Agency." The Daily Beast. March 19, 2021. (https://www.thedailybeast.com/nasa-veterans-baffled-by-reported-biden-pick-to-lead-agency)

Feldscher, Jacqueline with Bryan Bender. "Bolden Would Have 'Preferred' to See a Woman Lead NASA." Politico Space. *Politico*. March 26, 2021. (https://www.politico.com/newsletters/politico-space/2021/03/26/bolden-would-have-preferred-to-see-a-woman-lead-nasa-492250)

Smith, Marcia. "Nelson Greeted with Accolades at Nomination Hearing." SpacePolicyOnline.com. April 21, 2021. (https://spacepolicyonline.com/news/nelson-greeted-with-accolades-at-nomination-hearing/)

U.S. Senate Committee on Commerce, Science, & Transportation. "Nomination Hearing to Consider the Presidential Nominations of Bill Nelson to Be National Aeronautics and Space Administration Administrator. . . ." April 21, 2021. (https://www.commerce.senate.gov/2021/4/nomination-hearing)

Bartels, Meghan. "Biden Proposes $24.7 Billion NASA Budget in 2022 to Support Moon Exploration and More." Space.com. April 9, 2021. (https://www.space.com/biden-nasa-2022-budget-request)

Gohd, Chelsea. "NASA to Land 1st Person of Color on the Moon with Artemis Program." Space.com. April 9, 2021. (https://www.space.com/nasa-sending-first-person-of-color-to-moon-artemis)

Sadek, Nicole. "NASA Wants $11 Billion in Infrastructure Bill for Moon Landing." Bloomberg Government. June 15, 2021. (https://about.bgov.com/news/nasa-wants-11-billion-in-infrastructure-bill-for-moon-landing/)

Foust, Jeff. "Nelson Remains Confident Regarding Funding for Artemis." *SpaceNews*. October 3, 2021. (https://spacenews.com/nelson-remains-confident-on-nasa-funding-for-artemis/)

Foust, Jeff. "Revised Budget Reconciliation Package Reduces NASA Infrastructure Funds." *SpaceNews*. October 29, 2021. (https://spacenews.com/revised-budget-reconciliation-package-reduces-nasa-infrastructure-funds/)

Davenport, Christian. "Citing China Threat, NASA Says Moon Landing Now Will Come in 2025." *The Washington Post*. November 9, 2021. (https://www.washingtonpost.com/technology/2021/11/09/nasa-moon-artemis-spacex-china/)

NASA Office of Inspector General, Office of Audits. "NASA's Management of the Artemis Missions." Report No. IG-22-003. NASA. November 15, 2021. (https://oig.nasa.gov/docs/IG-22-003.pdf)

Press Release. "As Artemis Moves Forward, NASA Picks SpaceX to Land Next Americans on Moon." Release 21-042. NASA. April 16, 2021. (https://www.nasa.gov/press-release/as-artemis-moves-forward-nasa-picks-spacex-to-land-next-americans-on-moon)

Shepardson, David. "U.S. Watchdog Rejects Blue Origin Protest Over NASA Lunar Contract." Reuters. July 30, 2021. (https://www.reuters.com/business/aerospace-defense/us-agency-denies-blue-origin-protest-over-nasa-lunar-lander-contract-2021-07-30/)

Roulette, Joey. "Jeff Bezos' Blue Origin Sues NASA, Escalating Its Fight for a Moon Lander Contract." The Verge. August 16, 2021. (https://www.theverge.com/2021/8/16/22623022/jeff-bezos-blue-origin-sue-nasa-lawsuit-hls-lunar-lander)

Roulette, Joey. "Blue Origin Loses Legal Fight Over SpaceX's NASA Moon Contract." *The New York Times*. November 4, 2021. Updated November 10, 2021. (https://www.nytimes.com/2021/11/04/science/blue-origin-nasa-spacex-moon-contract.html)

Jewett, Rachel. "Senate Appropriations Directs NASA to Pursue Second HLS With $100M." Via Satellite. October 19, 2021. (https://www.satellitetoday.com/space-exploration/2021/10/19/senate-appropriations-directs-nasa-to-pursue-second-hls-with-100m/)

Brown, Mike. "Blue Origin New Glenn Specs, Power, and Launch Date for Ambitious Rocket." Inverse. July 23, 2021. (https://www.inverse.com/innovation/blue-origin-new-glenn-specs-power-launch-date-for-ambitious-rocket)

Bergin, Chris. "Amid Ship 20 Test Success, Starbase Prepares Future Starships." NasaSpaceflight.com. November 14, 2021. (https://www.nasaspaceflight.com/2021/11/ship-20-success-prepares-future-starships/)

Bender, Maddie. "SpaceX's Starship Could Rocket-Boost Research in Space." *Scientific American*. September 16, 2021. (https://www.scientificamerican.com/article/spacexs-starship-could-rocket-boost-research-in-space/)

Ferguson, Sarah. "Destination Mars." *Foreign Correspondent*. ABC (Australian Broadcasting Corporation) News. September 30, 2021. Updated November 1, 2021. (https://www.abc.net.au/foreign/destination-mars/13565384)

Gross, Jenny. "Satellite Monitoring of Emissions from Countries and Companies 'Changes Everything,' Al Gore Says." *The New York Times*. November 3, 2021. Updated November 6, 2021. (https://www.nytimes.com/2021/11/03/climate/al-gore-cop26.html)

"Climate Change: How Do We Know?" *Global Climate Change: Vital Signs of the Planet*. NASA. (https://climate.nasa.gov/evidence/)

Voosen, Paul. "NASA's New Fleet of Satellites Will Offer Insights into the Wild Cards of Climate Change." Science. May 5, 2021. (https://www.science.org/news/2021/05/nasas-new-fleet-satellites-will-offer-insights-wild-cards-climate-change)

Freedman, Andrew. "Al Gore's Climate TRACE Finds Vast Undercounts of Emissions." Energy & Environment. Axios. September 16, 2021. (https://www.axios.com/global-carbon-emissions-inventory-surprises-cb7f220a-6dfd-4f88-9349-5c9ffa0817e9.html)

"'The First Earthrise' Apollo 8 Astronaut Bill Anders Recalls the First Mission to the Moon." The Museum of Flight. December 20, 2008. (https://www.museumofflight.org/News/2267/quotthe-first-earthrisequot-apollo-8-astronaut-bill-anders-recalls-the-first)

"John F. Kennedy Moon Speech—Rice Stadium." September 12, 1962. (https://er.jsc.nasa.gov/seh/ricetalk.htm)

Sagan, Carl, "Pale Blue Dot: A Vision of the Human Future in Space," Random House, 1994.

EPILOGUE

Smithberger, Mandy, and William Hartung. "Demilitarizing Our Democracy." TomDispatch. January 28, 2021. (https://tomdispatch.com/demilitarizing-our-democracy/)

Public Citizen. "If you're spending . . ." @Public_Citizen Tweet. Twitter. January 7, 2021. 1:29 pm. (https://twitter.com/public_citizen/status/1347248913464532992?lang=en)

INDEX

ABOUT THE AUTHOR

LORI GARVER is a leading figure in the US space program who pioneered innovations that are transforming NASA. She was the principal advisor on aerospace issues to three presidential candidates and led the NASA transition team for President Obama. Garver served as Deputy Administrator of the space agency from 2009 to 2013 and is known as an architect of the new era of commercial partnerships that allow SpaceX to carry astronauts to and from the International Space Station.

Garver is the Founder of Earthrise Alliance, a philanthropic initiative utilizing satellite data to address climate change and cofounder of the Brooke Owens Fellowship, an internship and mentorship program for collegiate women interested in pursuing space careers. She serves as an Executive in Residence at Bessemer Venture Partners and a Senior Fellow at Harvard Kennedy School's Belfer Center.

Garver is the recipient of the 2020 Lifetime Achievement Award for Women in Aerospace and has been awarded three NASA Distinguished Service Medals. She lives in Washington, DC.